Biological globalisation

Bio-invasions and their impacts on
nature, the economy and public health

Biological globalisation

Bio-invasions and their impacts on nature, the economy and public health

Wouter van der Weijden
Centre for Agriculture and Environment

Rob Leewis
National Museum of Natural History Naturalis

Pieter Bol
Delft University of Technology

And although there is a Law of the Conservation of Matter, there is

no Law of the Conservation of Species.

Charles Elton (1958)

Invasions are important, invasions are interesting,

invasions are inadequately understood. (..)

Deliberate releases of any sort should be discouraged.

Mark Williamson (1997)

Preface

The world is globalising, and nature is no exception. More and more plants, animals and microbes are crossing natural barriers and settling in new areas. The sum total of these biological invasions can be termed 'biological globalisation'. Although this proliferation is anything but new, it is now accelerating as a result of climate change, the globalisation of traffic and transport and the breakdown of distribution barriers. One of the impacts of bio-globalisation is the decline in differences between biotas: 'biotic homogenisation'.

Charles Elton's groundbreaking *The Ecology of Invasions by Plants and Animals* was published in 1958. Since then several more books on the subject of bio-invasions have seen the light of day, and yet there are still very few books that describe not only the effects on nature and biodiversity, agriculture and other productive sectors, but also the consequences for public health. For the Netherlands there is no general overview available at all.

In 2003 it occurred to one of us, Wouter van der Weijden of the Centre for Agriculture and Environment, to write such a report. He then approached Pieter Bol at Delft University of Technology and Rob Leewis, who at that time worked at the Netherlands Environmental Assessment Agency. This led to a unique and fruitful collaboration between three scientists from diverse disciplines: a biologist specialising in zoogeography, terrestrial ecology and agriculture; an aquatic biologist; and a medical epidemiologist with a broad knowledge of history. The Dutch Ministry of Agriculture, Nature and Food Quality, which was already developing a policy paper on invasive exotic species, was willing to pay some of the costs.

The result was the report *Biologische globalisering – omvang, oorzaken, gevolgen, handelingsperspectieven* (Biological globalisation – extent, causes, effects and options, 2005). The next logical step was this English language book. KNNV Publishers agreed to publish the book and the Ministry of Agriculture, Nature and Food Quality made an additional financial contribution.
This unique book integrates biological, agricultural, medical, historical and economic knowledge on biological globalisation. It treats invading micro-organisms, plants and animals alike in their multiple impacts on nature and biodiversity, the economy and human health.

While writing this book, we became completely spellbound by the diversity and complexity of bio-invasions and their impact. There were four reasons for that.

First, though we all know that every species is unique, bio-invasions are even more unique. Every single invasion reflects a unique combination of a species, a way of dispersal and crossing barriers, a subsequent dispersal pattern, new ecological interactions and sometimes an impact on nature and society. No two invasions are exactly the same. Every invasion has a unique history and a story to be told.

Second, there are also common patterns, which although still poorly understood are slowly being unravelled. They are just as fascinating as the differences and operate on a global scale. Like biogeography, already a global science, ecology too is now going global.

Third, many bio-invasions are the outcome as well as the cause of interactions between nature and societies. Societies accidentally or intentionally help species cross barriers, and then these species may have an impact on the local ecology, and thereby on society, ranging from nothing to anything between a blessing and a disaster. Few people realise how important the role of bio-invasions has been in history. Bio-invasions bring biologists, historians, agronomists, medical scientists and other disciplines together.

Finally, bio-invasions are important for the future of life on earth. Their impact on biodiversity is more often negative than positive, at least from a global perspective. They contribute to more homogeneous, less diverse and less stable biotas. In that respect they are not much less important than the well-known causes of habitat loss, climate change and exploitation, but this is insufficiently reflected in the media and in the policies of governments and NGOs.

With this book we want to convey to our readers some of our fascination for the subject. We also hope to draw more attention to an underestimated element in the ongoing debate on globalisation. We have written this book for all those working or studying in multiple scientific disciplines, policy-making, business and nature conservation, as well as for the general reader with an interest in bio-invasions and their multiple impacts, both in the Netherlands and elsewhere.

This book basically tells the same story in two tracks: the main text and the boxes. The main text provides systematic information and the theoretical background; the boxes mainly contain examples and cases and do not run strictly parallel to the main text. The boxes can be read separately.

Welcome to the fascinating world of biological invasions!

Wouter van der Weijden MSc, biologist
Rob Leewis PhD, aquatic biologist
Pieter Bol MD PhD, medical epidemiologist

Contents

Introduction

More and more barriers in the world are disappearing. As a consequence, more and more interchange of goods, services, information, money and people is taking place between regions, leading to countless new interactions. This process, known as globalisation, has been the source of much debate over the past decades. Although the discussion has been primarily concerned with economic and socio-cultural globalisation, there is also a third form of globalisation that has increasing and multiple impacts and is causing increasing debate: biological globalisation.[1] This is the process by which biological species spread across classical geographical distribution barriers, leading to so-called bio-invasions. These bio-invasions can be subdivided into two distinct categories: natural and anthropogenic invasions.

Natural invasions are caused by intrinsic changes in a species (genetic or epigenetic changes) or by natural changes in the environment (for example a natural climate change followed by sea level change). The main natural change in the climate over the last 10.000 years has been the end of the Ice Age, which was followed by countless northward bio-invasions. A recent example of an intrinsic change in a species is the expansion of the fulmar *Fulmarus glacialis* since 1800 from Iceland across the Atlantic as far as Brittany; and in the opposite direction: the expansion of the collared dove *Streptopelia decaocto* since 1930 from Asia and the Balkan across all of Europe (Williamson 1997).[2]

By contrast, *anthropogenic invasions* are facilitated by humans. This can be done in various ways:

1 *Intentional introductions of species from the opposite side of a geographical border.* This involves species for hunting, fishing or production (agriculture, forestry, livestock husbandry, aquaculture), as well as for pets and ornamentals. Occasionally a species is *re*-introduced that had gone extinct locally. Introduced species may or may not spread spontaneously. If they don't spread, many authors do not consider them "invaders".

2 *Unintentional introduction of species by transport across a geographical barrier.* Usually this is associated with the four T's of globalisation: Trade, Transport, Travel and Tourism. It often includes small species such as rats, insects and plants with small seeds.

3 *Invasions caused by the breakdown of a geographical barrier* between two separated bioregions. For instance: the digging of a canal connecting two seas, or the building of a road through a continuous forest belt. In both cases species can invade in two directions.

4 *Invasions caused by changes in land use and habitat.* For instance, the clearing of a forest for agriculture will attract weeds. And the building of houses will attract cockroaches and the house sparrow *Passer domesticus.* Such invasions often do not include crossing a geographical barrier.

5 *Invasions caused by climate change*, enabling species to survive and spread in areas where they could not before. Again, this often does not include crossing a geographical barrier.

Note that not every introduction leads to an invasion (in the stricter sense), whereas not every invasion is caused by an introduction. Many combinations of the six categories do occur. In fact the same species may figure in two or three of these categories. For instance, the settlement and expansion of *Aedes* mosquitoes in North America in the 1990s is partly due to transport from Asia and partly due to climate change.

This book mainly deals with invasions across geographical barriers, so the categories 4 and 5 will receive less attention. Th emphasis will be on category 2: transport- and traffic-related invasions. This category is most closely associated with the globalisation of society.

When did anthropogenic biological globalisation start? The earliest starting point that can be taken is between 200,000 and 100,000 years ago, when *Homo sapiens* left Africa; but generally, much more modern starting points are being used (see Appendix 1).

The oldest written source on bio-globalisation is a letter of Darius the Great of Persia (521-468 BC) ordering the warden of one of his estates to introduce "trees and plants on the other side of the Euphrates" (Bright 1998). In the 1st century AD, Pliny mentions two species of trees introduced in Rome: the plane tree (sycamore) from Greece and the balm (balsam) tree from Palestine. In addition, he mentions a number of animals from Africa and Asia that were imported to perform in circus games: the African and Indian elephants, lion, leopard (panther), tiger, monkeys, rhinoceros, giraffe and crocodile.[3] He also mentions the unintentional introduction of leprosy from Egypt and of the skin disease *lichen* from Asia Minor.

Marco Polo reported that Kublai Khan, the Mongol ruler who conquered eastern China in AD 1279, searched for evergreen trees and, whenever he came across such trees, would have them dug out and - using elephants if necessary - transferred to his arboretum in Beijing (Bright 1998). The early 15th century saw the first journeys of exploration from Europe but simulta-

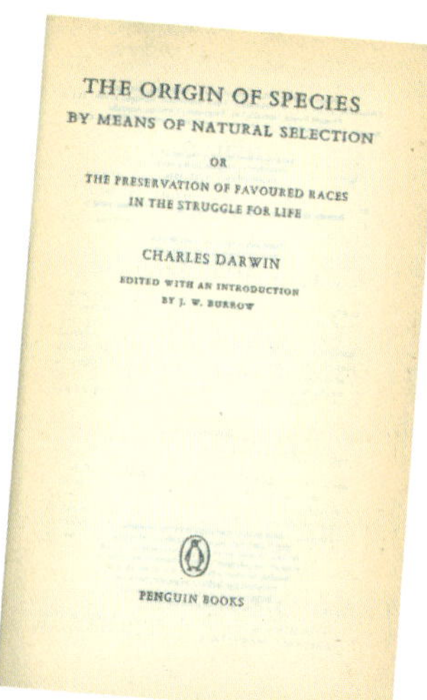

Figure 1.

Three milestones in the history of invasion biology, first published in 1859, 1958 and 1996, respectively. All three have British authors.

neously also from China (see Box 7). Since that time, the number of deliberate and accidental introductions on all continents has grown rapidly.

Although bio-invasions have a long history, invasion *biology* is still a young branch of science. Bio-invasions are treated briefly by Lyell in *Principles of Geology* (1832) and by Darwin in *The Voyage of the Beagle* (1836). Bio-invasions were one of the inspirations for Darwin in formulating his theory of evolution. *On the Origin of Species* (1859) repeatedly offered solid facts and arguments relating to bio-invasions; together, these arguments can even be seen as the foundation for invasion biology (Ludson & Wolfe 2001).

However, the study of bio-invasions remained anecdotal until 1958. In that year, Charles Elton published *The Ecology of Invasions by Animals and Plants*, laying the foundation for a systematic approach and thereby raising invasion biology to a higher level.[4] A more recent standard work is *Biological Invasions* by Mark Williamson (1997). Since 1999 there is even a scientific periodical *Biological Invasions*.[5] However, as Vermeij (1996) remarks in a brilliant article, the formation of theories and empirical testing are still at an early stage. Some confusion even remains about such elementary terms as "invasive" (Colautti & MacIsaac 2004). Experimental research has only recently taken off (J. Memmott cited in Van Strien 2004). In medical epidemiology, however, the study of invasions has been underway for a much longer time.[6]

As with all forms of globalisation, bio-globalisation offers both advantages and disadvantages to nature and the human society. In this book we will consider the following questions:

- What is the nature and scope of biological globalisation?
- What are the drivers, mechanisms and pathways of distribution?
- What patterns can be observed in bio-invasions?
- What are the impacts on nature, environment, the economy and public health?
- What is the position and role of the Netherlands in bio-globalisation?
- What impacts do bio-invasions have in the Netherlands?
- Can we, and should we control bio-invasions?
- What needs to be done?

2

The nature and scope of bio-globalisation

The role of geographical barriers in evolution

Although many species have developed strategies for crossing geographical barriers (see Box 1), such barriers have played an important part in the history of life on Earth. A barrier is an area with a habitat that is unsuitable to particular species – an area that those species cannot cross. The term "barrier" is, of course, relative: what constitutes a barrier to one species can serve as a perfect bridge to another. Obviously, land is a barrier to most aquatic organisms; land as well as salt water is a barrier to most freshwater organisms; and land, salt water and freshwater are barriers to most estuary-based species.[7] A high mountain range acts as a barrier to most lowland organisms and a desert is a barrier to species requiring a humid habitat.

Species differ greatly in ecological amplitude and dispersion ability. Many species are confined to a small range of ecological conditions. Many birds, by contrast, can fly over deserts and seas. The eel *Anguilla anguilla* migrates from the Sargasso Sea across the Atlantic Ocean and North Sea into freshwater inland waterways; it even travels across land.

Crucially, even relatively small barriers can be effective, e.g. the Panama isthmus, the Andes mountain range and the Mozambique Channel. The (partially) deep sea between Bali and Lombok even marks an important division between two zoogeographical regions, Wallace's line. Yet it is no more than 30 km in width, and was even narrower during the Ice Ages.[8] In the Amazon, a river of a few hundred meters' width can be an effective barrier to land-based species, even those that are able to swim, because caimans and predatory fish prevent them from crossing.

As soon as a distributional barrier has formed somewhere, divergent evolution starts to occur. The older and more effective a barrier, the larger the taxonomic differences tend to be. Initially, endemic species are formed; millions of years later, whole endemic genera, families and orders can develop. A crucial event in the history of life on Earth was the disintegration of the mega-continent of Pangaea, 180 million years ago, forming major dispersion barriers to land species, including the Atlantic

Ocean and the Turgai Strait (between Europe and Asia). From that moment on, the biotas of America, Europe, Asia, South America, North America and India followed separate evolutionary pathways.

About 33 million years ago, however, the Turgai Strait disappeared, and Europe was reconnected with Asia (Reemer 2005). This lead to a major biotic interchange as well as many extinctions, called *La Grande Coupure* by scientists. Between 3 and 3.5 million years ago, another major reconnection took place: the Panama Isthmus was uplifted and reconnected South America with North America. That was the starting point of the so-called Great American Interchange. This included cats, bears, dogs, horses and rabbits invading South America; and conversely, monkeys, opossums and capybaras invading North America. Many species went extinct. Meanwhile a third major reconnection was realized, although intermittently: North America and Asia were connected repeatedly when Beringia (the Bering Strait) emerged from the sea. This again caused several biotic interchanges, including the three-toed horse *Hipparion* spreading from America to Europe (about 11 million years ago) and *Homo sapiens* spreading from Asia to America (over 14,000 years ago).[9]

In recent geological times, however, the Beringian interchange was confined to the Ice Ages, when sea levels were low and only species of cold and temperate climates could cross. Between the (sub) tropical biotas of the Old and New World there was hardly any interchange, so they continued to evolve separately.

Islands and lakes

Islands are a story on their own. They usually harbour less species than do comparable areas of equal size on the mainland. We can separate islands into two categories: continental islands and oceanic islands.

Continental islands are often located on a continental plateau; they had a land connection to the mainland during the Ice Ages, allowing species interchange between the island and the mainland. On such islands, the British Isles for instance, we consequently find very few or no endemic species. The fauna is often a more or less impoverished version of the fauna on the adjacent mainland.

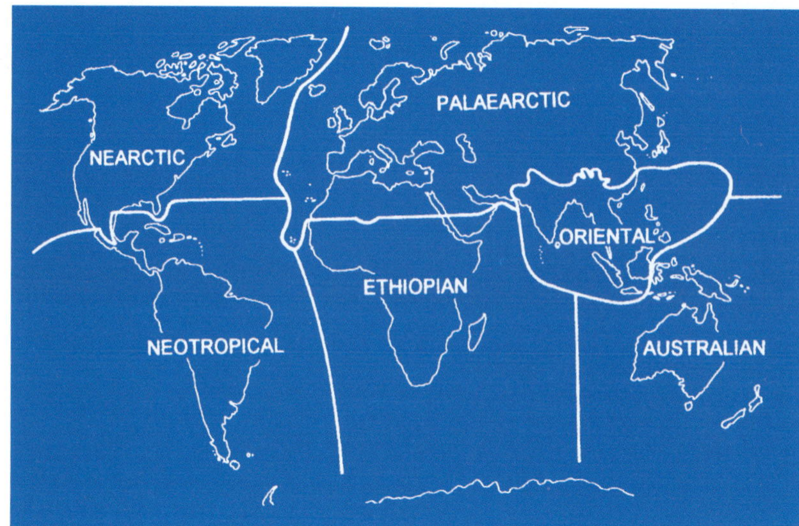

Figure 2.
The classical zoo-geographical regions as described by P.L. Sclater and slightly adapted by Wallace (1876) (After Williamson 1997). Note that none of the regions exactly matches a continent.

Oceanic islands, by contrast, have never had such a land bridge to the mainland. They are volcanic islands, coral islands or the remains of an older continent (for example Madagascar or the Great Antilles) or a mini continent (New Zealand). The volcanic and coral islands are populated entirely with species that managed to cross the sea from the mainland or from other islands, either on their own (see Box 1) or by hitching a ride. The taxonomic composition of the fauna on such islands is not representative. We often find birds, plants, insects, spiders and reptiles, but few or no amphibians (which are tied to freshwater) and few or no land mammals. Even New Zealand did not harbour any land mammals except bats until the arrival of Polynesians in the 10th century AD. Land predators, even carnivorous ants, are often entirely absent. An extreme example of poor and limited flora and fauna is the remote Easter Island in the Pacific. Here only 44 native plant species were found in 1917, next to a mere 5 native species of land animals (Elton 1958).[10]

Although oceanic islands have a relatively small number of species, they often have a relatively large number of *endemic* species, particularly large and remote islands like those in the Pacific Ocean, as well as the Canary Islands. A striking feature of some remote archipelagos is that multiple species occur which are morphologically quite different but taxonomically closely related. The underlying process is called adaptive radiation: a burst of evolution with rapid divergence from a single ancestral species that results from an array of habitats and niches. The Galapagos finches ("Darwin's finches") are the most famous example, but the honeycreepers of Hawaii are even more spectacular (see chapter 5). They co-evolved with the Lobelia species that also underwent adaptive radiation.

Lakes are the mirror image of islands. Similar processes occur here. However, whilst islands are being "fed" by the species pool of the nearest continent, most lakes are not (or rarely) fed by seas and oceans. In this aspect, lakes can best be compared with remote islands, though many lakes do have connections to other bodies of water.

A population that becomes isolated will evolve in two different ways. On the one hand the species will continue to adapt to its specific environment. On the other hand, the species can lose characteristics that are no longer needed. This is most evident on oceanic islands. Here, many species have lost much of their defence mechanisms against land-based predators, parasites and diseases. They have become "ecologically naive". Some birds have even lost their ability to fly. Famous examples are the dodo on Mauritius and the kiwi and the kokako in New Zealand.[11]

In addition, islands often harbour miniature or giant versions of species, such as the dwarf cassowary on New Guinea, the giant earworm on St. Helena and giant locusts (*weta*) in New Zealand. In the past, large mammals often evolved in miniature forms on islands, such as the dwarf mammoth on Wrangel Island in the Arctic Ocean, which became extinct only 4000 years ago. Another example is the recently discovered *Homo floresienses*, who did not become extinct until the last ice age. By contrast, small animals sometimes develop giant forms on islands, like the giant rat and the now extinct elephant bird *Aepyornis maximus* (up to 3m tall) of Madagascar.[12]

Through this mechanism of divergent evolution on islands and continents, geographical barriers have contributed greatly to the biodiversity of the world.

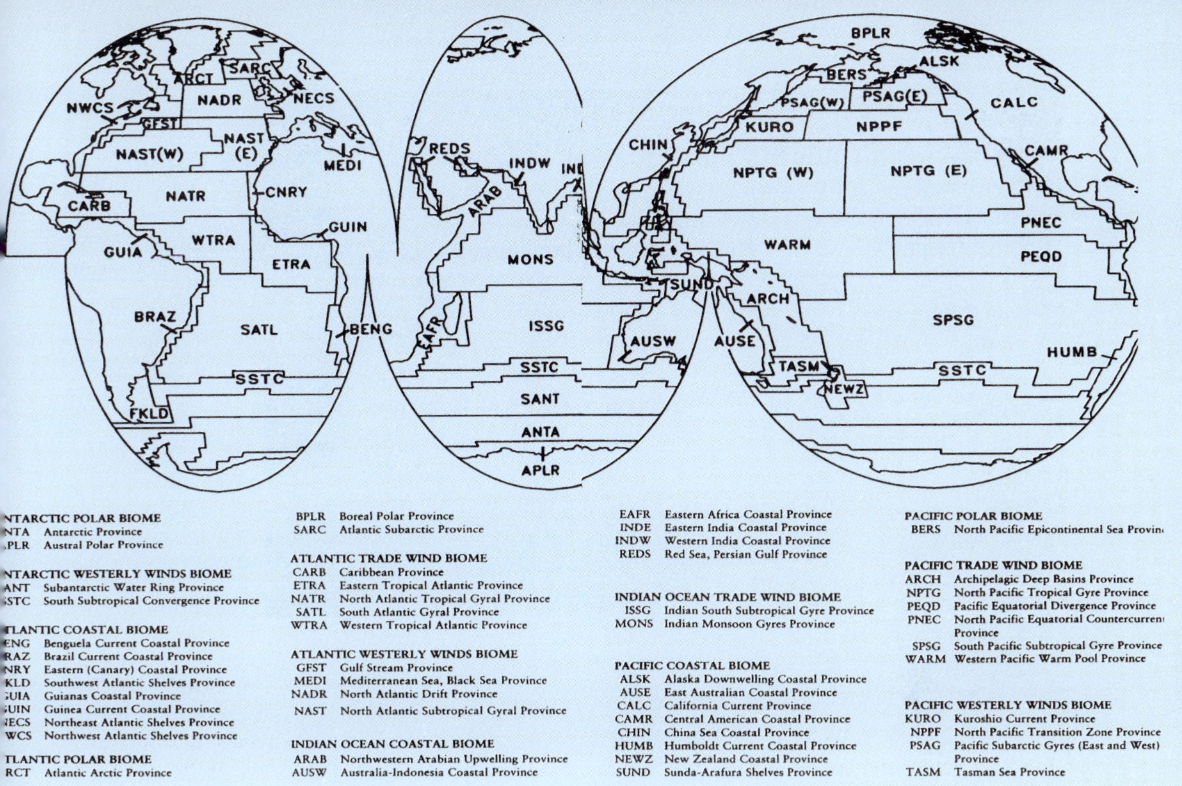

ANTARCTIC POLAR BIOME
ANTA   Antarctic Province
APLR   Austral Polar Province

ANTARCTIC WESTERLY WINDS BIOME
SANT   Subantarctic Water Ring Province
SSTC   South Subtropical Convergence Province

ATLANTIC COASTAL BIOME
BENG   Benguela Current Coastal Province
BRAZ   Brazil Current Coastal Province
CNRY   Eastern (Canary) Coastal Province
FKLD   Southwest Atlantic Shelves Province
GUIA   Guianas Coastal Province
GUIN   Guinea Current Coastal Province
NECS   Northeast Atlantic Shelves Province
NWCS   Northwest Atlantic Shelves Province

ATLANTIC POLAR BIOME
ARCT   Atlantic Arctic Province

BPLR   Boreal Polar Province
SARC   Atlantic Subarctic Province

ATLANTIC TRADE WIND BIOME
CARB   Caribbean Province
ETRA   Eastern Tropical Atlantic Province
NATR   North Atlantic Tropical Gyral Province
SATL   South Atlantic Gyral Province
WTRA   Western Tropical Atlantic Province

ATLANTIC WESTERLY WINDS BIOME
GFST   Gulf Stream Province
MEDI   Mediterranean Sea, Black Sea Province
NADR   North Atlantic Drift Province
NAST   North Atlantic Subtropical Gyral Province

INDIAN OCEAN COASTAL BIOME
ARAB   Northwestern Arabian Upwelling Province
AUSW   Australia-Indonesia Coastal Province

EAFR   Eastern Africa Coastal Province
INDE   Eastern India Coastal Province
INDW   Western India Coastal Province
REDS   Red Sea, Persian Gulf Province

INDIAN OCEAN TRADE WIND BIOME
ISSG   Indian South Subtropical Gyre Province
MONS   Indian Monsoon Gyres Province

PACIFIC COASTAL BIOME
ALSK   Alaska Downwelling Coastal Province
AUSE   East Australian Coastal Province
CALC   California Current Province
CAMR   Central American Coastal Province
CHIN   China Sea Coastal Province
HUMB   Humboldt Current Coastal Province
NEWZ   New Zealand Coastal Province
SUND   Sunda-Arafura Shelves Province

PACIFIC POLAR BIOME
BERS   North Pacific Epicontinental Sea Province

PACIFIC TRADE WIND BIOME
ARCH   Archipelagic Deep Basins Province
NPTG   North Pacific Tropical Gyre Province
PEQD   Pacific Equatorial Divergence Province
PNEC   North Pacific Equatorial Countercurrent
       Province
SPSG   South Pacific Subtropical Gyre Province
WARM   Western Pacific Warm Pool Province

PACIFIC WESTERLY WINDS BIOME
KURO   Kuroshio Current Province
NPPF   North Pacific Transition Zone Province
PSAG   Pacific Subarctic Gyres (East and West)
       Province
TASM   Tasman Sea Province

Figure 3.
Marine biogeographical regions according to Longhurst (1998).

Effects of crossing geographical barriers

We humans are causing geographical distribution barriers to be increasingly crossed or even destroyed. This results in bio-introductions, sometimes followed by bio-invasions, with two different possible effects on biodiversity.

The first possible effect is that a species, after entering into new territory, occupies a vacant niche; or the species and indigenous species redistribute existing niches, and manage to co-exist. Biodiversity in this case does not decline globally and even increases locally.

The second possible effect is that indigenous species become extinct through predation, grazing, parasitism, disease or competition. Biodiversity then remains stable regionally but declines globally. If an *endemic* species becomes extinct, the loss is irreparable.

Bio-invasions have been reported in all land, freshwater, and marine ecosystems, both in cultivated and natural areas, and even in Antarctica.

Some examples of the increasing share of exotics in *terrestrial* systems on islands and continents:

- On many oceanic islands goats, pigs, deer, monkeys, dogs, cats, mongooses, rats, snakes, snails, ants, parasites and pathogens have been introduced, often causing great harm to the native fauna. As Darwin already remarked in 1859: "Hardly an island can be named on which our smaller quadrupeds have not become naturalised and greatly multiplied".

- Iceland permanently harbours six exotic terrestrial mammals, in addition to the single indigenous mammal (Arctic fox): the house mouse, field mouse, brown rat, black rat, reindeer and mink. Several of these have become pests for humans; at least one, the mink *Mustela vison*, has developed into a pest for wild birds (Génsbøl & Feilberg 2003).

- On Hawaii, of the 7828 insect species found, 2582 (33%) are alien. Of the 2690 plant species, 946 (35%) are alien.

- On the equally isolated but much smaller Easter Island, the five native species of animals counted in 1917 were already far outnumbered by 49 introduced species (including one that probably had settled spontaneously), a ratio of almost 1:10! (Elton 1958).

- The average number (over the past 150 years) of plant species establishing on New Zealand was

Country/region	Native species	Introduced species	Percentage
Antigua/ Barbuda	900	180	17%
Australia	15,000-20,000	1500-2000	9%
Austria	3000	300	9%
Canada	3160	881	22%
Ecuador - Rio Palenque	1100	175	14%
Finland	1250	120	9%
France	4400	500	10%
Guadeloupe	1668	149	8%
Hawaii	1200-1300	228	15%-16%
Java	4598	313	6%
New Zealand	1790	1570	47%
Spain	4900	750	13%
US	17,000	5000	23%

Source: Heywood (1989). US: Pimentel (ed) (2002)

12 per year (Williams & Timmins 2002).

- The percentage of introduced plant species in a selection of countries and regions varies from 6% on Java to 47% in New Zealand (Table 1).
- On continental islands, the number of exotic species is also growing fast. In the UK, for instance, 46% of the terrestrial mammals are exotic, as well as 10% of the birds, 33% of the reptiles, 57% of the amphibians and 23% of the fish species (White & Harris in: Pimentel 2002).

In the US, including Hawaii and Guam, introduced plant species constitute 23% of the total flora (see Table 1).[13] Among harmful species, the share of exotics can be even higher. In US pastures 45% of the weed species are exotics, and in agriculture even 73%. In New Zealand as much of 90% of harmful invertebrates are exotic (Cook *et al.* 2002).

Examples of the numbers of exotics in *aquatic* systems (Bright 1998):

- In the Baltic Sea the number of invasive species is 89 and in Australia it is 124.
- The Mediterranean Sea already contains 480 invasive species.
- In the San Francisco Bay Area, the number of successful invasions has risen from about one every *year* in the period 1851-1960 to one every *quarter* of a year in the period 1961-1995.[14] Locally the fauna even consists of 100% exotic species.
- Some freshwater canals in Florida and Hawaii are dominated by introduced aquarium fish species.
- In the US, the total number of exotic species of plants, animals and microbes is estimated to be 50,000 (Pimentel *et al.* 2005).

The process of bio-globalisation (or bio-proliferation) is well underway; the result is a worldwide homogenisation (or increasing entropy) of the flora and fauna.[15] As we shall see later on in this book, bio-invasions are occurring more and more frequently, both on land and in water, and the end of this process is nowhere in sight.

How plants, animals and microbes naturally cross barriers

Some plant and animal species are able to cross large geographical barriers, and settle in an area located thousands of kilometres away. They do this either actively or passively, or through a combination of the two. Even extremely remote islands have been colonized in this way. **Active distribution** occurs primarily among animals and takes place by walking, crawling, flying or swimming. **Passive distribution** means that the species makes use of the movement of rivers, sea currents, air or animals.

Some examples of passive distribution:

Sea currents can carry along icebergs, floating islands and trees for thousands of kilometres. This enables plants (seeds), reptiles and insects to "hitch a ride". Darwin already showed that many plant seeds keep their ability to germinate even after 50 to 100 days in salt-water. The crustacean *Calanus finmarchicus,* common in the Northern Atlantic, also occurs around Madagascar, probably using the cold northern currents that dive downwards and travel below the warmer surface waters of the tropics to South Africa and beyond (Elton 1958). Likewise, groundbreaking research by Rita Colwell and colleagues has shown that the cholera bacterium can travel enormous distances on and in copepods (miniscule crustaceans) carried along by ocean currents such as El Niño.

Air currents can transport microbes, small animals such as spiders, and plant seeds and spores over large distances, and they can help flying animals to bridge extra large distances. Storms and hurricanes, in particular, can carry plant seeds over large distances and force migrating birds off course. According to Wallace (1880) this mechanism has played a key role in the colonization of remote islands like the Azores and Bermudas.

Some **birds, whales, bats and butterflies** travel actively (sometimes passively) over distances of thousands of kilometres. Parasites and plant seeds and spores hitch a ride in their stomachs and bodies, or attached to their feet, feathers and hairs. Darwin showed that pieces of mud stuck to a bird's foot sometimes contain germinative seeds. In the same way, fish can end up in isolated lakes. Wallace (1880) noticed a remarkable similarity between the flora of southern and northern cold regions. He argued that this was, among others, because sea birds travel enormous distances with seeds and spores attached to their feathers.

Nevertheless, many species of plants and animals fail to cross even small barriers, such as narrow straits. But these species increasingly get a helping hand from humans, both intentionally and accidentally.

Means of transport as a vehicle in bio-invasions

Means of transport are playing an increasingly greater part in bio-invasions. In the past, these were pack animals, horse carriages and ships; more recently also trains, cars and airplanes.

Some examples of bio-invasions facilitated by various means of transport:

Pack animals:
- In the Middle Ages, the plague bacterium was spread across Asia by camels and horses, which carried clothes and furs that in turn carried infected fleas.

Ships:
- Plants, worms, beetles and other animals were introduced in soil and waste material brought as ballast by sailing ships.
- Plant pests and diseases as well as weeds are brought with shipments of plants and plant products including wood. For instance: the centuries-long import of wool for use by local industries (Bright 1998).
- Aquatic species are brought in ballast water and as fouling of ship's hulls. Some dramatic examples: the zebra mussel in America (see Box 52) and the ctenophore *Mnemiopsis* in the Black Sea (see Box 29).
- Species like the clinging jellyfish *Gonionemus vertens* and Japanese wire weed *Sargassum muticum* have been imported with transports of oysters for aquaculture.
- The plague bacterium was spread as early as the 6th century with black rats and their fleas on ships.
- European colonist crossing the Atlantic introduced a battery of pathogens (including smallpox, measles and influenza)
- European slavers introduced African diseases in America with slave ships (malaria tropica and yellow fever, both insect-borne diseases).

Trains:
- Trains spread many plant species along their tracks.

Cars:
- Tourists driving home from their summer holidays in Southern Europe spread plants, animals, parasites and pathogens.
- Anglers in the US spread aquatic species such as water fleas while cleaning their fishing equipment in another lake (MacIsaac *et al.* 2004).
- Trucks helped one of the worst forest plagues in the US, the gypsy moth *Lymantria dispar*, by spreading its caterpillars.
- Truckers and their prostitute clients spread AIDS in its early years (pre-1985) in Uganda, Rwanda and Kenya. The same phenomenon is currently occurring in Eastern Europe, where roadside prostitutes are engaging in unsafe sex with tourists and truckers.

Airplanes:
- Air transport of fresh vegetables, flowers and plants spread plant pathogens and parasites across the globe.
- Airplanes introduced the Western corn rootworm *Diabrotica virgifera* from America to the Balkan in the early 1990s. From there it spread spontaneously across Europe. Additional invasions from other airports followed.
- Bird trade by airplane may have introduced the West Nile virus from Africa or the Middle East in New York (Spielman & D'Antonio 2001). Another possible pathway is spreading by migratory birds. This invasion may have been facilitated by favourable conditions (such as a warmer climate and increased flooding) for the mosquitoes (particularly the Northern house mosquito *Culex pipiens*) that transmit the virus from birds to humans.*
- Air traffic probably spread the HIV virus to the US (California) in the early 1980s, presumably from Africa, where it may have originated from ape viruses (see also Box 63). From there the virus was spread further, again by air traffic. At least 100 of the initial 200 patients in the US had been infected by a single Canadian airline steward working on domestic flights ("Patient Zero").

*www.rivm.nl/infectieziektenbulletin/bul119/westnile-virus3.html

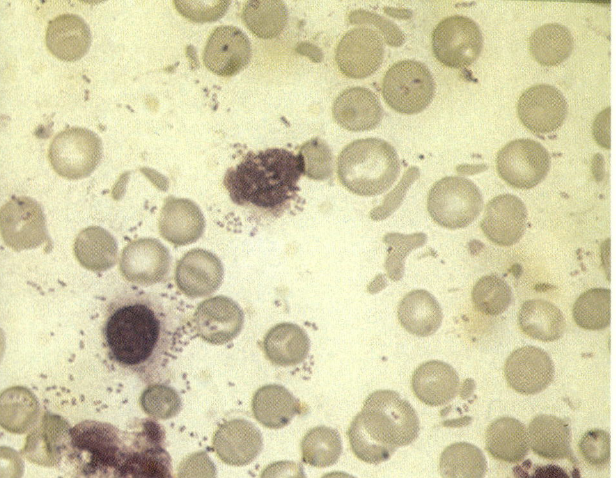

The plague bacterium among red blood cells.

Ship ballast:
playing ecological roulette

The use of ballast for sea ship stabilization is very old. Several materials were used, including stones, wood and soil. That explains the curious fact that many ancient Dutch fortresses in Brazil, Africa and Southern Asia were built using bricks from the Netherlands, as one can still observe. It also explains why near sea seaports large soil heaps were found, which made excellent bridgeheads for exotic plants.

After the invention of the electrical water pump, water became the usual ballast. This introduced risks of spreading aquatic species. Early warnings were put forward by Charles Elton in 1958.

Great tanker ships transport over 200,000 m² of ballast water. While pumping in the water, they also bring in large amounts of silt. Some of the water cannot even be pumped out.

In one analysis of 159 ships arriving from Japan in the Port of Coos bay in Oregon over a period of five years, 367 species were found, many of them exotics. These included 72 crustaceans, 43 annelid worms, 11 protozoa and 128 diatoms. Carlton & Geller characterized this kind of transport as "ecological roulette".

Ballast water has also been the source of invasions of cholera and insect-borne diseases.

Wood, too, is still being used as ballast, facilitating the spread of beetle species.

Sources: Elton (1958), Carlton & Geller (1993), Bright (1998) and Reumer (2005)

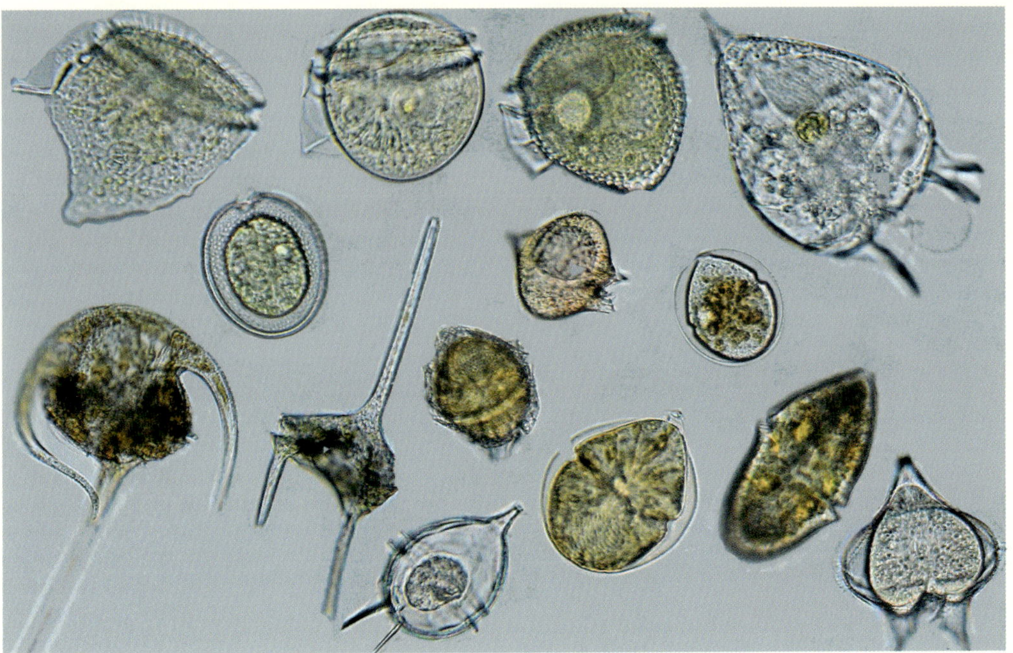

Dinoflagellate algae from Newfoundland and the Bay of Biscay found in ballast water discharged in the port of Rotterdam in 2006.

Transmission pathways of infectious diseases

Few transmission pathways can be ruled out when it comes to infectious diseases. In Box 2 we saw the plague bacterium travelling by ships and pack animals; a battery of pathogens crossing the Atlantic by ship; the HIV virus travelling by car; and the West Nile and HIV viruses travelling by airplane. In most cases animals (fleas, rats) or humans are carrying the pathogen. In some cases, ballast water or water containers carry the vector. Examples are cholera, malaria, yellow fever and dengue.

Pathogens can be introduced in association with other animals or products as well:

- *Salmonella* can be transmitted in contaminated **eggs** and **poultry**, which can also carry *Campylobacter*.
- *E. coli O157* can be transmitted in **beef** and in **plant products** like alfalfa; even cider can contain the bacterium.
- *Listeria monocytogenes* can be found in **soft cheeses**. These can also contain *Brucella melitensis*, a bacterium causing brucellosis (undulant fever or Malta fever).
- Imported **tomatoes** are sometimes infected with the *Pseudomonas* bacterium.
- Spores of the anthrax bacterium can even be transmitted by non-living objects called **fomites,** such as wool, wool products and hair for brushes.

One case that deserves special attention is ***Clostridium*** in **salmon**. It offers an illustration of the sneaky pathways a bacterium (in this case not an exotic) can follow to reproduce and spread.

Salmon can become infected with the extremely toxic *Clostridium botulinum*, which produces the botulin toxin. This is perhaps the most potent poison in nature, just a few milligrams being sufficient to wipe out an entire city population. It is invisible, tasteless and odourless. *Clostridium* thrives under anaerobic conditions.

Salmon is perfectly safe to eat once heated and canned in fish factories. Nevertheless, there were several fatalities in Birmingham in 1978 due to botulism. The victims had all eaten canned salmon from the same Alaskan brand. Investigation of the machine used to can the salmon showed that a tiny metal burr had caused a miniscule puncture in the cans. Presumably, the bacterium (of the dreaded type E) had penetrated through this puncture after the contents of the can had sufficiently cooled. The crews of the fishing boats were found to be the source. After returning from the sea, they tossed their wet oilskin jackets onto the warm cans in order to dry them. This allowed tiny drops of water containing *C. botulinum* to penetrate the cans. The warmth of the cans stimulated the bacterium to reproduce rapidly. The perforation did not allow enough oxygen inside the can to block growth, but it did prevent the cans from bulging – normally a clear sign of infection, even to a layperson.

This is how some unsuspecting people in Britain fell victim to a bacterium from the Pacific Ocean and a minuscule piece of metal 10,000 kilometres away.

Source for Clostridium: *Ball* et al. *(1979)*

The plague bacterium:
a mass killer and a history-maker

Few invasive species have had such a profound impact on the history of humankind as the plague bacterium. It generated three pandemics of increasing scale, each lasting from one to four centuries.

The bacterium *Yersinia pestis* originates in two regions: Central Africa and southwest Arabia, where it has old reservoirs in burrowing rodents. A third old reservoir may have occurred in India on the southern slopes of the Himalays. Fleas can transmit the bacterium to other rodents, such as the black rat *Rattus rattus,* though the rats become ill and often die, and therefore cannot function as a reservoir. Fleas can also survive for weeks on a diet of grains but they need blood to reproduce.

The global career of the bacterium was associated with the increasing shipping of grains and with the expansion of the black rat from India associated with sea ships. In the Middle Ages, the bacterium also found a permanent reservoir in the burrowing rodents of the Eurasian steppes. The bacterium is transmitted not only by rat fleas (bubonic plague) but sometimes also directly from humans to humans by air droplets spread by sneezing and coughing (pneumonic plague). Bubonic plague has a case-fatality rate of 70-90%, pneumonic plague of up to 100%.

Physician wearing protective clothes during the Black Death period, including glasses to protect the eyes.

First pandemic AD 541–810

The first pandemic started by AD 541 as bubonic plague in Egypt and Libya, which had been invaded from one of the three reservoirs. From Egypt rats and their bacteria were spread by ships and troops throughout the Mediterranean and the Middle East. In Constantinople (Istanbul), the capital of the East Roman (Byzantine) Empire, about 300,000 people reportedly were killed in AD 542. Even the emperor Justinian became ill though he recovered; hence the name "Plague of Justinian". The plague returned in waves until 750. From AD 610 to 806, the bacterium hit China, causing a decline in population size; the same happened in Japan around 808.

The political impact of this pandemic was enormous:

- Emperor Justinian had ambitious plans to reunite the East Roman and West Roman Empires and his troops had entered Italy. These plans were completely frustrated.
- When the Muslim expansion from Arabia started in 642, the resistance of the Asian and European peoples was already weakened.
- Within Europe, the centre of power shifted from the Mediterranean region to the north, where the black rat and its deadly parasite did not occur or was rare.

Second pandemic AD 1346–c.1740

The second plague pandemic had a different origin, spread more widely and lasted even longer than the first. Some scholars claim it originated in the steppes of East Asia, others assume its origin was in the steppes of Western Asia and Eastern Europe. Conditions for the spread of the disease were favourable since the Mongols were at the peak of their power in AD 1279–1350, ruling an area from China in the east to the Ukraine west and south into Iraq and Iran. They spread infected fleas to the west, enabling the bacterium to establish new reservoirs in burrowing rodents.

By 1346 Mongols converted to Islam brought infected fleas to the Crimea peninsula, where they were trying to expel Christian Italians from their fortified trading colonies. Italian ships subsequently brought infected fleas and rats to Constantinople and several Mediterranean ports including Messina (on Sicily) and Marseille. From there *Yersinia pestis* started its devastating European career. A shipping network developed during the previous centuries included Northern Europe, so this time the rats, their fleas and their bacteria could quickly spread far north. The Black Death caused the largest ever mortality rates in Europe, with 25-60% of all Europeans succumbing.

In the following centuries, waves of the plague contin-

Black rat

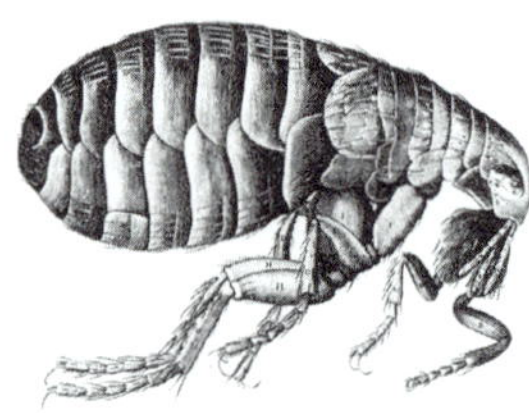

Rat flea

ued to strike Europe at intervals until the disease began to decline after c. 1680, with occasional though sometimes severe local outbreaks well into the 18th century. This decline was hastened by changes in house construction as well as a new bio-invasion. The black rat often climbed into thatch roofs, from where their fleas would fall onto humans. This happened less often from the mid 17th century onwards as thatch roofs were gradually replaced by stone roofs, which provided less suitable conditions for the rat. The late 18th century saw an invasion of the brown rat *Rattus norvegicus*, which gradually displaced the black rat. Although the brown rat can carry the bacterium as well, it is a wilder species that does not live as close to humans and it infested roofs and house walls less often, preferring to burrow in the ground.

The impact of the second pandemic was, again, tremendous:

- The world population shrank noticeably (see Figure 4).
- At least in Europe, whole areas were depopulated and forest cover invaded agricultural land. Some of these so-called *Wüstungen* still exist today.
- The high and repeated mortality created a dramatic shortage of labour. As a result, wages increased and land rents fell structurally. This meant a huge transfer of income from the property-owning class to the working class, including women. It also accelerated the development of a labour market and, more generally, a market economy.
- The forest re-growth that followed the first waves of the plague may even have contributed to a cooling of the climate (see Box 42).
- In the 17th century, the plague hit Spain particularly hard, contributing to the decline of this superpower. Among the nations that benefited was the Dutch Republic, which by the time was revolting against the Spanish occupation (1568-1648) while simultaneously developing into a maritime superpower (see Box 72). The Republic itself also suffered from plague outbreaks

during the war, on average one every two years, but most of these were local and less devastating than those in Spain.

Third pandemic AD 1855–1959

The third pandemic was a really global pandemic and lasted a century. It started in Yunnan, southern China, during a Muslim rebellion in the 1850s. Chinese troops returning home spread infected fleas to the north. In the 1880s, the plague reached Chinese seaports, including Canton and Hong Kong. Benefiting from the steamship network that had developed in the 1870s, the rats and their bacteria spread to India, Africa, South America, North America and Australia within a decade.

Remarkably, the bacterium established new reservoirs in burrowing rodents in South Africa, Argentina and western North America; in the last case the arrival of the brown rat in 1899 played a key role. By 1940 the bacterium was found in 34 species of rodents and 35 species of fleas worldwide. But then, an effective drug had been discovered: antibiotics. In 1959 the WHO officially declared the pandemic to be over. Its scale had been larger than ever, but the mortality rate and political impact had been limited. This time, Europe had been hit only briefly.

In addition to being a mass killer and a history-maker, the plague was a maker of literature, inspiring such famous writers as Boccaccio (Black Death in Florence 1348), Samuel Pepys and Daniel Defoe (Great Plague of London 1665) and Albert Camus (fictional plague outbreak in the Algerian city of Oran in the 1940s). Their fascination mainly regarded the disruptive impact the plague had on human relationships.

Sources: Benedictow (2006), McNeill (1998), Flinkenflögel (1992), Childs Kohn (2001), Bryan (1999) and Noordegraaf & Valk (1988)

First account: Boccacio on the Black Death in Florence 1348

The second pandemic of the plague hit Europe from 1347 onwards. In the next year, Giovanni Boccaccio gives a first account of the Black Death in Florence in the introduction to his famous *Decameron* (1349-1353):

"......And it did not behave as it did in the Orient, where if blood began to rush out the nose it was a manifest sign of inevitable death [*pneumonic plague*]; but rather it began with swellings in the groin and armpit, in both men and women, some of which were as big as apples and some of which were shaped like eggs......[*bubonic plague*]

...But what gave this pestilence particularly severe force was that whenever the diseased mixed with healthy people, like a fire through dry grass or oil it would rush upon the healthy...

...enormous amounts of refuse and manure were removed from the city by appointed officials, the sick were barred from entering the city, and many instructions were given to preserve health..."

Source: Boccaccio (transl. G.H. McWilliam 1972)

The explosion of explorations in the 15th century

For thousands of years the system of Silk Roads formed the backbone of Eurasian trade. It was the main trade route in the world. Precious goods were traded westward and gold and silver eastward. The East and the West had important common interests in keeping this arterial route open.

However, from the beginning of the last millennium the system came under increasing pressure from the South as the Ottoman Empire expanded. In Europe it was feared that once the system was in Turkish hands, a monopoly and high prices would be the result. This was an additional reason for the Crusades. Diplomatic contact between East and West intensified. Travellers like Marco Polo, Carpini and Rubroek were not just adventure seekers, they were looking for possibilities to forge alliances with the Mongol Empire and China (some of them even had mandates in that direction). An alliance was never forged, though.

While Europe engaged in conflicts in the Middle East, China, under the new rule of the Ming Dynasty (1368–1644), made a surprising move. At the beginning of the 15th century it launched several long expeditions across the oceans with huge fleets. Obviously, there were economic reasons for this twist in the traditional attitudes of the 'Empire of the Middle'. A substantial part of the planet was soon mapped.* The Chinese voyagers collected many animals and plants and some authors claim that the Chinese brought sweet potato, taro, yam and even maize to Southeast Asia.

In the same period Portugal started to send fleets southward. It was a long process, but eventually they reached South Africa and India in 1498. Did they meet Chinese fleets? No, they had vanished. After some successful voyages, the Chinese emperor Cheng Zu had suddenly grasped that these ventures would completely change both the world at large and internal relations in China. He also had to defend the northern borders of his country against ongoing attacks from the Mongols. By 1435 he ended the explorations and ordered that all visible traces, including objects, maps and scriptures, be destroyed.**

In 1453 Constantinople fell into Turkish hands and so did the traffic on the Silk Roads, with the feared consequences for the flow of goods and money. This heightened the zeal for exploration in the West. Spain joined in, followed by other European nations during the 16th century. Within a few decades the worldwide human web had developed and the global market was created. That was the start of the ever-increasing global interchange of living material: animals, plants, seeds and microbes.

Biological globalisation as a phenomenon is not uniquely linked to mankind, but the human factor has predominated in the last few thousand years. In this process, the 15th century was a true turning point.

* Menzies' (2003) recent claim that the Chinese had already mapped all the world's oceans by 1421 has been criticised by several historians.

** Only now, 570 years later, has the sleeping giant China woken up, with all the global and internal consequences.

Sources: Duyvendak (1938 & 1949), Muller (1944), Wallerstein (1976) and Levathes (1994)

History, drivers, mechanisms and pathways of bio-invasions

Early history

Anthropogenic bio-globalisation began between 100.000 and 200.000 years ago, when *Homo sapiens* spread from Africa to Eurasia. Distribution took place over both land and sea. Migrants carried with them – both deliberately and accidentally – plants, seeds, parasites and microbes. Later, dogs were domesticated and joined the party. Following the invention of agriculture, probably first in the Middle East about 10,000 years ago, crops, seeds and livestock began to be actively introduced. By that time, the cat was domesticated in the Middle East. It helped protect stored grain against rodents.

Heydays of trade and immigration were the age of the early Silk Roads (at the time of the Roman empire), the age of the Chinese Tang dynasty and Persian Abbasid dynasty (AD 600-900), and the age of the Mongol empire (13th and 14th century). Bio-invasions followed not just the Silk Roads through Asia, but also caravan routes through the Arabian Peninsula and the Sahara.

Crossing the seas, humans had already reached Australia some 40,000 years ago, followed by Oceania. Some 14,000 years ago, but perhaps much earlier, humans arrived in North America via the Bering Strait and subsequently colonized South America. There was some transoceanic interchange of crop species. For example, Indonesian sailors introduced taro and yams in Madagascar around AD 500 or even earlier (McNeill & McNeill 2003). Sweet potato was introduced from South America to Oceania, but how this happened is not yet clear.

During the Middle Ages, the Atlantic was repeatedly crossed by Icelanders, the western population branch of the Vikings. The first colonization attempt was as early as AD 998-1002. The Icelanders named the newly discovered land *Vinland*.[16] These attempts probably led to the interchange of several marine species (Haydar pers. comm.) although just one documented case of this is known: the soft-shell clam *Mya arenaria* introduced in the Western Atlantic (Petersen *et al.* 1992). But there was little interchange of terrestrial species. First of all because the attempts did not last long, due in part to attacks from Amerindians. Secondly, relatively few invasions by *new* species were possible that far north, because the

northern biotas of the Old and New World are relatively poor in species and already shared many species due to earlier biotic interchanges, mainly via the Bering Strait.

Sweet potato travelled from South America to Southeast Asia before Columbus, but how that happened is still a mystery.

The Columbian Exchange

In the 15th century the Chinese as well as the Portuguese made long-distance ocean voyages (see Box 7). The great breakthrough came with the "discovery" of America by Columbus in 1492 and the subsequent successful colonizations. Columbus himself started the interchange already at his second voyage in 1493, when he tried to introduce into Hispaniola a battery of crops: chickpeas, fruit stones, grape vines, melons, onions, radishes, salad greens, sugar cane and wheat. He added a variety of animals: cattle, chickens, dogs, goats, horses and sheep.

The *conquistadores* that followed Columbus did not travel alone but were joined by their pathogens, horses, war dogs and pigs (Crosby 1972). Conversely, maize arrived in Spain as early as AD 1494 and quickly found its way to the rest of Europe, Africa and Asia. Those were the starting points of a growing list of species introductions in both directions, breaking through the geographical barrier of the Atlantic, which had formed after the disintegration of the Pangaea super-continent. For the first time in 180 million years the biotas of tropical and subtropical America and Eurasia were interconnected.[17] This process, which included microbes, plants, animals as well as humans, has become known as the Columbian Exchange (Crosby 1972).

Remarkably, pathogenic organisms travelled almost exclusively one-way, from the Old World to the New World. Europeans and African slaves, as well as their ships, brought with them a battery of communicable diseases to America (see Box 8). They literally decimated the Amerindian population. Only one pathogen doubtlessly travelled in the opposite direction: the sand flea, which is no serious pathogen by itself, but may create wounds that pave the way for more serious diseases such as tetanus. Syphilis, a much more important disease, possibly is also of New World origin (see Box 11).

Livestock species, too, travelled mainly one-way. Only two species were successfully introduced from the New World into the Old World: the turkey and the Muscovy duck. Humans travelled one way as well, from Europe and Africa to the Americas. The same holds for weeds (see Box 8).

By contrast, the interchange of crop species was almost symmetric. This interchange made a major contribution to the growth of global food production, since New World crops could be grown on soils that, prior to 1492, were rated as useless because of their sandiness, altitude, aridity or other factors. Conversely, Old World livestock made the unproductive American grasslands a rich source of protein. The accelerated growth rate of the world population after 1650 (which was to continue until approximately 1970) would have been impossible without the Columbian Exchange of crops and livestock (Crosby 1972, McNeill & McNeill 2003).

Another factor that contributed much to biological globalisation was the increasing number of long-distance ship voyages as well as the increasing speed of ships,

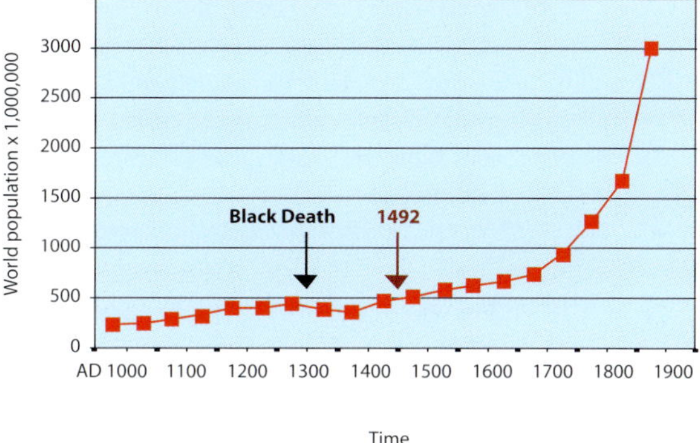

Figure 4.
Growth of the world population AD 1000-1950 showing the effect of the Black Death (Calculated from estimates by Biraben 1980 and McEvedy *et al.* [1978]; 1900 and 1950: UN 1999).

allowing more living plants, animals and microbes onboard to survive long enough to be introduced. For the survival of plants onboard, the invention of the so-called Wardian Case (see Box 14) around 1830 made a remarkable contribution. A next milestone in bio-globalisation was the emergence of a steamship network in the 1870s. And of course, the introduction of the airplane in the early 20th century was another milestone.[18]

Drivers

Which are the basic drivers of biological globalisation?

The main driver is life itself. Everybody who has a garden knows that every species tends to reproduce itself, and spread. Without that tendency, bio-invasions simply would not occur.

It would be absurd, however, to hold nature responsible for the rapid increase of invasions over the last centuries and the last decades. Human society itself is driving the process in various ways:

- The human species itself tends to reproduce and spread rapidly. In fact we are the most invasive species in the history of the planet.
- Humans have directly introduced, both intentionally and accidentally, a great number species in all parts of the world.
- Population growth leads to increased use of

natural resources and increased consumption, production, traffic, transport, climate change, and exploitation of habitats. All these factors tend to enhance bio-invasions. Habitat destruction breaks down distribution barriers and creates the disturbed habitats that so many invaders prefer.

- Technological developments in traffic and transport allow increasingly faster movements in increasingly larger amounts over increasingly greater distances. The greater speed also provides organisms with greater chances of survival en route.

- Economic growth per capita reinforces all these effects.

- Political developments further enhance the process. These include international pressure to remove trade barriers (often exerted by export oriented countries, multinational corporations and multilateral organisations such as the World Trade Organisation, the International Monetary Fund and the World Bank). National governments promote trade and traffic through various measures such as creating transport infrastructure and tax exemptions for commercial shipping and air traffic.

These drivers of biological globalisation are largely the same as the drivers of economic and socio-cultural globalisation.

Mechanisms and pathways

The introduction of species can take place in four different ways.

Intentional introduction into the environment takes place for specific purposes, in particular food production (agriculture and aquaculture). In aquatic environments, fish and shellfish species are deliberately introduced for cultivation. On land, crops, livestock and game (e.g. the rabbit, originally from the Iberian peninsula) were introduced in this way.

Sometimes a species is *intentionally introduced for local use, after which it is accidentally introduced into nature.* Examples on land are the muskrat in Europe and the rabbit in Australia. Examples in water are the red king crab *Paralithodes camtschatica* in Norway and the Japanese (or Pacific) oyster *Crassostrea gigas* in the Netherlands. In 1983 wastewater from the Oceanographic Museum in Monaco was dumped into the Mediterranean. At that time, no one suspected that within 20 years this action would lead to a global infestation of the seaweed *Caulerpa taxifolia* (see Box 15).

Accidental introduction also takes place. Species may "hitch a ride" with humans, plants, animals or means of transportation. On land, mice, rats and plant seeds are distributed in this way. Ballast water and the biological fouling of ship's hulls are vectors for many aquatic organisms. Seaweeds and parasites may sneak in with oysters imported for aquaculture. Microbes, parasites and weeds are accidentally introduced both on land and in water.

Figure 5.
The Suez Canal connected the separate biotas of the Red Sea and the Mediterranean Sea, leading to a large number of bio-invasions, particularly into the Mediterranean.

The *breakdown of a distribution barrier*, too, may result in accidental introductions. The barrier involved can be a land barrier between two bodies of water; or, less often, a water barrier between two landmasses, for instance an island and a nearby continent.

The first major breakdown of a *land* barrier was the construction of the Great Canal between the Yangtze River and Yellow River in China in AD 611. In 1859 the Suez Canal was completed, connecting the Red Sea with the Mediterranean Sea.[19] The Panama Canal joined the Atlantic and Pacific Oceans in 1914, though this did not lead to species interchange because the canal is at a relatively high altitude and fed by rivers. In 1992, the Rhine-Main-Danube Canal connected the Rhine and Danube river basins, creating a freshwater connection between the North Sea and the Black Sea.

Breakdown of a *water* barrier usually involves shallow waters. One example is the Meuse-Scheldt estuary in the province of Zeeland in the Netherlands; the islands here have been connected with the mainland by dams. Species interchange in such cases will be rare, since the islands involved are continental islands with similar fauna and flora.

Numerous pathways have been identified for invasions, varying from air traffic to ballast water and trade in plants, livestock and pets. Appendix 5 provides a near-complete overview. The importance of the various pathways differs from one taxon to another. In the US, for instance, most introductions of plants and vertebrate animals were deliberate, while most introductions of invertebrates and microbes were accidental (Pimentel *et al.* 2005).

Success of invasions

Far from every settlement of an organism is successful. In fact, most settlement attempts fail after a brief or more prolonged period of time. In America, for instance, numerous species of birds from the Old World have settled, including thrushes, finches, tits, the nightingale, woodlark, European robin, dipper, corncrake, capercaillie and mute swan (Elton 1958). Most however disappeared after a few years. Only six species of birds have managed to settle permanently: the house sparrow *Passer domesticus*, starling *Sturnus vulgaris*, common partridge *Perdix perdix* and pheasant *Phasianus colchicus* from Europe, as well as *Phasianus torquatus* and the crested mynah *Aethiopsar cristatus* from Asia. All are species that inhabit an urbanized or agrarian environment.

Even after a successful invasion, a species (or variety) can become extinct. This, for example, happens every year following a global invasion of one or more new strains of influenza.

An imported species can reach several stages of penetration:

1 Import of the species into a closed-off environment (cage, aquarium, enclosure etc.).
2 Introduction of the species into the wild (by escaping or being released).
3 Settlement: the species is able to maintain itself and reproduce.
4 Development of a species into a pest.

Roughly speaking, every next step has a 10% chance of success. This rule of thumb is known as the "tens rule" formulated by Williamson (1997). So: out of every 1000

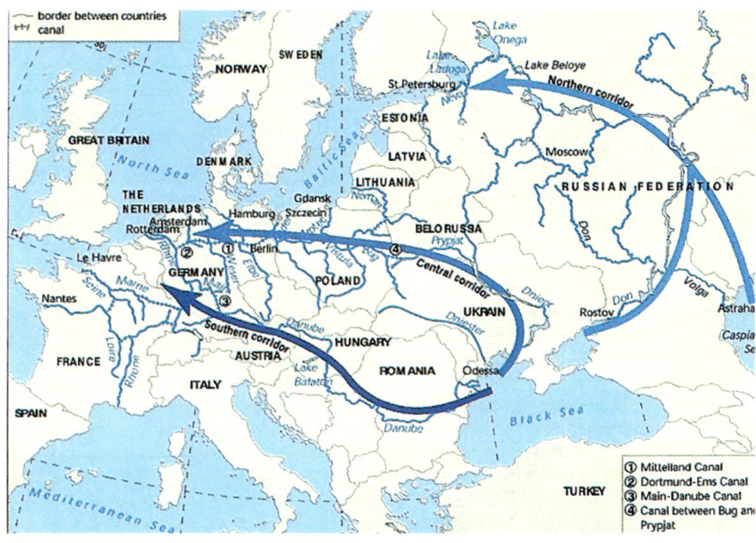

Figure 6.

Several canals in Europe helped Ponto-Caspian aquatic species spread north and west (From: Bij de Vaate *et al.* 2002).

imported species, 1000 x 0.1 x 0.1 x 0.1 = 1 species develops into a pest. See Table 2 for an example.

Problem is that the terms "pest" and "serious pest" are not clearly defined, and they are a hybrid of biological, economic and subjective terms. This makes it impossible to test the validity of this part of the tens rule. Williamson himself lists four exceptions to the rule – cases in which the percentage of success is significantly higher than 10%.[20] However, we find many other higher, and very few lower percentages in the literature. For example[21], out of 80 species of exotic vertebrate animals in Australia, 30 (37%) are considered pests, particularly in agriculture (Bomford & Hart 2002).

In an attempt to formulate a more robust rule, Jeschke & Strayer (2005) differentiate between three more easily testable steps: 1) introduction; 2) maintenance; and 3) spreading. They calculated the success rate of each of these steps for the introduction of vertebrates (primarily fish, birds and mammals) from Europe to North America, and vice versa. Their figures indicate that step 1 is the least successful: 8% of European vertebrates have been introduced in North America, while 4% of North American vertebrates have been introduced in Europe. For step 2, the rate of success is much greater: 60% for introductions from Europe to North America and 52% for introductions in the opposite direction; and the success rate for step 3 is at least as high: 65% and 54%, respectively.

The same authors also compiled figures from nine other publications, varying in scope from flatworms and insects to birds and mammals, and from islands and continents. Without exception the success rates exceeded 10%, in some cases even reaching 100%. This corresponds to figures on the damage and benefits of the 97 introduced birds in the US. Of these, 5% are considered beneficial, while most (56%) are pests (Temple cited in Pimentel *et al.* 2005).

Jeschke & Strayer do not exclude the possibility that the tens rule does apply to plants. But even for the 5000 alien plant species in the US, it has been claimed that *most* have displaced several native plants (Morse *et al.* cited in Pimentel *et al.* 2005)

In other words, there do not appear to be any generic rules; at best, specific rules can be applied. The tens rule remains an underestimate for rates of success, at least for introductions of animals. That is a fact of relevance to policy-makers.

Also relevant to policy: often it is assumed that introduced agricultural crops or ornamental plants cannot successfully spread from the plots where they have been sown or planted because they would not survive in other biotopes. However, some cases of this have been reported (see Box 23). Nowadays the trade in ornamental plants even is the key source of plant invasions.

The Columbian Exchange, and its asymmetry

Unlike the voyages of the Icelanders around AD 1000, the first voyage of Columbus to America in 1492 was the starting point of a massive transatlantic flow of people, goods, plants, animals and microbes. Crosby (1972), focusing on crops, livestock, humans and their infectious diseases, called this the "Columbian Exchange".

A striking feature of the exchange is that it was far from symmetric. Many more livestock species, infectious diseases and weeds travelled from the Old World to the New World than vice versa. The same holds for humans: Europeans and their African slaves far outnumbered the very few Amerindians that travelled or were brought to Europe.

Crops, by contrast, show a much more symmetric pattern. Vavilov (cited in Crosby 1972) made up a list of 640 most important plants domesticated by man. Roughly speaking, 500 of them belonged to the Old World (including 50 from Africa) and 100 to the New World. But if we focus on the successful crops, the exchange was much more symmetric, since the New World list includes such important crops as maize, potato, sweet potato and cassava.

How can we explain those asymmetries as well as the exception of crops? Probably they have a common root. In Eurasia before AD 1500, livestock husbandry as well as city networks had developed much further than in America. These two factors created a wide range of pathogens and diseases, a pan-Eurasian "civilised disease pool", which in turn induced widespread immunity among people against a battery of diseases. Thus, at the start of the Columbian Exchange the Europeans had a much wider range of livestock species, diseases and immunities. They could beat the Amerindians not primarily because of their weapons and horses, but because they had powerful allies: pathogens. Starting with smallpox and measles, these pathogens killed the Amerindian population on a scale without precedent. The Aztecs were decimated as early as 1568 and the entire Amerindian population by 1602.

That created a major labour shortage, which made the Portuguese and Spanish decide to introduce slaves from Africa. The slave ships first introduced the most dangerous malaria parasite, *Plasmodium falciparum*, which found competent vectors among indigenous mosquito species, then the yellow fever virus and its vector *Aedes aegypti*. These pathogens killed even more Amerindians (as well as Europeans), accelerating the replacement process.

Of course, the human pathogens cannot explain the asymmetry in the exchange of weed species. This may have been the result of the Europeans introducing their cultivation systems into the New World (see Box 9).

Many other species were involved in the exchange, including pets, ornamental plants, feral plant and animal species as well as parasites and pathogens of crops and livestock. For instance, the lama and alpaca populations in South America were decimated not only by brutal exploitation by the Europeans, but also by diseases transmitted by introduced livestock.

The Columbian Exchange still continues. New species are crossing the Atlantic, such as the corn rootworm that is currently invading Europe. But since 1900, the interchange is more or less symmetric. For vertebrates, the direction has even been reversed.

Sources: Crosby (1972 & 2000), McNeill (1998), Diamond (1999), Bryan (1999), McNeill & McNeill (2003) and Jeschke & Strayer (2005)

Modern replica of the *Santa Maria*, Columbus' flagship in 1492, in Barcelona.

The Columbian Exchange of crop species, livestock species and infection diseases

	From Old to New World	From New to Old World
Crop species	banana barley black pepper buckwheat cabbage citrus coconut coffee egg-plant (aubergine) garlic lentil lettuce nutmeg oats olive onion pea peach pear rice rye sorghum soy spelt sugar beet sugar cane turnip wheat	avocado beans cashew cassava chilli pepper kina (cinchona) cocoa cranberry maize (corn) papaya paprika peanut pecan pineapple potato rubber squash (incl. pumpkin) sunflower sweet potato* tobacco** tomato vanilla
Livestock species	cattle donkey fowl (incl. chicken) goat horse pig rabbit sheep	Muscovy duck turkey
Infectious diseases**	cholera influenza measles plague smallpox tropical malaria tuberculosis typhoid yellow fever	sand flea (chigger) ? syphilis (venereal type)

** Introduced into Oceania and Southeast Asia before 1492*
*** Smoking is the most devastating "disease" introduced from the New to the Old World. But of course, smoking is neither a disease nor a species. It is a habit that can cause several deadly diseases. Smoking started invading Europe in 1493, before the tobacco plant itself was introduced in 1531 (Borio 2006).*

Europe as an ecological imperialist

Eurasia has a highly positive "species interchange balance": it has provided many more invasive species to the rest of the world than vice versa. As much as 80% of introduced plants on other continents is of Eurasian origin (Di Castri 1989). This asymmetry can be observed in both terrestrial systems (Crosby 2000) and aquatic systems (Leppäkoski *et al.* 2002). The reasons are economical and political rather than biological.

At first sight the asymmetry is surprising. One might expect European ships to have sailed as often *to* as *from* other parts of the world. This, however, is not the case. Up until the 19th century, ships sailed to other continents more often than they sailed back to Europe, sometimes because they sank or were hijacked, more often because they remained in Asia and other regions to serve in regional traffic. For example, the Dutch East India Company (VOC, AD 1602-1799) organised nearly 5000 ship passages *to* and just 3000 *from* the Far East (see Box 72). This explains some of the asymmetry.

A more important factor, at least for terrestrial species, may be that Europe colonized other regions, not vice versa. The colonists introduced European environments: cultivation systems characterized by intensive soil and grassland exploitation as well as European-style cities, gardens, game, ornamental plants and even songbirds. Thus Europe became an "ecological imperialists", and created many "Neo-Europes" in other temperate zone regions in North America, Argentina, South Africa, Australia and New Zealand (Crosby 1986).

Two additional factors have played a role:
1. A ship often remains moored in harbour much longer in the country where it originates and/or was constructed than in a country where it sails for trading. Therefore, a species in the country of origin has a better chance of colonizing the hulls (e.g. shellfish) or cargo space (e.g. rats) of the ship than a species in the destination country. Hence ships sailing from Europe were more "loaded" with species.
2. European ships imported more cargo (including oil, since the previous century) than they exported. Conversely, they exported more ballast than they imported. Ballast often contains more species than does cargo. Stones, soil, wood and water were used, all with associated species. Even today, a major asymmetry in the ballast water balance occurs in the Netherlands (see Box 76).

The asymmetry has decreased over time with the end of colonialism and the rise of the US as an economic super power. It may even be reversed, now that many Asiatic economies are growing rapidly, and many sea-ships are being built there.

First account: syphilis 1493

In 1497 Nicolo Leoniceno was the first to describe the syphilis epidemic that had begun in 1493 (*Libellus de Epidemia, quam vulgo morbum Gallicum vocant*):

"......now a disease of an unusual nature has invaded Italy & many other regions. In the beginning pustules are on the private parts, soon on the whole body and frequently located on the face itself besides causing great hideousness as well as a great deal of pain. Moreover to this disease the physicians of our time do not yet give a name, but is called by the common name of French disease, as if this contagion were imported from France into Italy or because Italy was invaded at the same time both by the disease itself & the armies of the French."

Source: Major (1965)

Syphilis: origin and impact

The origin of syphilis has long been, and still is being disputed among experts. One complication is that the disease is caused by the same spirochete bacterium *Treponema pallidum* that causes yaws, a disease already endemic in Europe before 1492. However, whilst many pre-Columbian palaeo-pathological traces of syphilis have been found in America, hardly any such trace has been found in the Old World. Possibly, syphilis was spread by skin contact in America, but in Europe - where skin contact was limited by clothing habits – rapidly adapted into a sexually transmitted disease. The spread may have been enhanced by the flourishing bathhouse culture.

Possibly, syphilis arrived in Europe when Columbus returned from his first voyage to America in 1493. The pilot of the *Santa Maria* had caught the disease and spread it further. A big opportunity for the bacterium came soon: after the siege of Naples in 1494-1495, international troops returning home accelerated the spread of the disease. People often named the disease after the most despised neighbouring country: "Spanish pox", "German disease", "English disease", "Polish disease", and so on. Soon after, the Turks would call it "Christian disease" and the Persians "Turkish disease" (see Box 11). Tragically, Columbus himself died from syphilis.

The virulence of the disease contributed to the chastity offensive in the centuries that followed. For example, bathhouses were closed almost throughout Europe. Syphilis may even have helped the spread of the chastity-preaching Islam and Christianity in Southeast Asia.

The 19th and 20th centuries saw syphilis becoming less virulent in Europe as well as Asia. This, it is hoped, will also happen over time to that recently emerged sexually transmitted disease: HIV/AIDS.

Sources: Canter Cremers - van der Does (1965), Crosby (1972), Roberts (2003) and Boomgaard (2006)

Albrecht Dürer's The Syphilitic (1496).

Introductions facilitated by discoverers, horticultural gardens and acclimatisation societies

Voyages of discovery have made a significant contribution to biological globalisation. For instance, the British biologist Joseph Banks, who accompanied Cook on his travels in the 18th century, collected plants from everywhere and brought them back to Britain, where he became the director of Kew Gardens. Banks and his staff collected some 300 species of plants (or plant seeds), as well as fish, birds and insects, on Tahiti alone. Many of the plants were cultivated further in Britain. Conversely, he planted a large number of seeds on Tahiti (water melons, oranges, lemons etc.) that he had brought with him from Rio de Janeiro. On July 4th 1769, he proudly wrote of this in his diary. Darwin (1836) found 225 species of higher plants on the Galapagos Islands, and brought back 193. He did not mention whether they included living plants and seeds.

The first **botanical gardens** were founded by Europeans as early as the 16th century, both in their own countries and in their colonies. Perhaps the first was the botanical garden of the University of Cologne (1490-1516). Around 1545 gardens were started in Pisa, Padua, Florence, Bologna, Zürich, Kassel, Leipzig, Montpellier and many other cities with universities. Among the cities that followed were Valencia (1567), Leiden (1594), Amsterdam (1638) and Utrecht (1639). In 1694 a *Hortus Medicus* was founded in South Africa by the Dutch East India Company (VOC) after the model of the Amsterdam garden.

One of the goals was to allow plants from other parts of the world to become acclimatized, so that they could be planted out for medical, economic or ornamental goals. Colonists wanted plants from their native countries. Conversely, people in their home countries (including colonists who had returned back home) wished to enrich their gardens and landscapes with exotic species.

Economic introductions included rubber trees from South America in the Malay Archipelago, coffee plants from Africa in South America, and banana plants from Southeast Asia in Central America. Species introduced in Europe included trees grown for wood production.

Many ornamental plants became invasive. For example, of the 3150 exotic woody plants currently grown in German gardens, at least 210 have become invasive in Central Europe. A global survey of 1060 woody plant invasions found that in the 624 cases in which the origin of the invasion could be ascertained, 59% came from botanical gardens, landscaping or other amenity purposes.

European carp, introduced on a large scale in the US.

A global network of **acclimatisation societies** developed after 1850.
The American society made multiple unsuccessful attempts to introduce thrushes, finches, skylarks and nightingales. Success was achieved only with the introduction of the starling in 1890, which quickly spread from New York City, reaching the West Coast within 50 years.

Many attempt were also made to introduce hunting game and fur animals, including deer, chamois and pheasants. These species often proved to be harmful to the indigenous flora.

In 1871 the US Fish and Fisheries Commission was founded. The Commission started to introduce various species of fish in the US. The common or European carp *Cyprinus carpio* in particular was introduced on a large scale, with active cooperation from the railways.

In Australia no less than six acclimatization societies were founded, which introduced fish, birds, mammals and plants from the Old World as well as the New World. Woodpeckers were introduced to control parasites of the eucalyptus, and secretary birds to control snakes. Rabbits, deer, lamas and pheasants were introduced as hunting game. Most introductions failed, but some were successful: the starling, sparrow, deer, rabbit and prickly pear cactus. But the latter two proved disastrous.

Source: National Geographic Magazine (November 1996), Bright (1998) and Werger (1989)

Botanical Garden in Groningen (the Netherlands).

<table><tr><td></td><td>

Wearisome plant introductions in the Cape Province
</td></tr></table>

The deliberate introduction of a species in a new environment is often a tiresome experience, as it has been demonstrated by many attempts at many places at many times. Although the number of introductions is increasing worldwide, *successful* introductions are relatively scarce.

This is clearly illustrated by attempts made by Jan van Riebeeck, founder of the Cape Colony under the Dutch East India Company in 1652, to introduce crops. To understand the problems he faced, one should know that the Cape has a Mediterranean climate, with winter rains, unlike most of sub-Saharan Africa and most of Europe. Consequently, common African and European crops could not be grown in the Cape. Until 1652, agriculture was even virtually absent, being limited to sporadic, primitive cultivation by local groups like the Khoi or Khoikhoi, and the San, called Bushmen by the colonists. This changed when, from 1652 onward, the Dutch started introducing citrus and other plants.

Right from the start, Van Riebeeck had to cope with the climate. He began his attempts in the South-African autumn and winter, using Dutch plants and seeds that were "expecting" spring and summer. His journal reads:

"On the 12th of June 1652. In the evening there were many loud peals of thunder and much lightning, with very heavy windstorms which at times came with such terrific violence that we thought everything would be blown down. All our young cabbage and many other Dutch vegetables and fruits, which were thriving beautifully, have been blown to bits and completely ruined".

"23[d] of July 1652. (the weather having calmed down somewhat) found all our hard work done in the garden completely flooded and all our crops submerged and spoilt. It was very sad to behold, as we had several beds sown and planted with wheat, barley, peas, cabbage, and other field and garden crops, some of which had grown so well that they were a pleasure to see. There had been such a heavy downpour that at several places the land looked like a sea, as the rivers could not hold all the water".

However, the Dutch didn't give up and after many efforts managed to introduce the whole range of their familiar horticultural crops. Sometimes it's dogged as does it.

Source: H.B. Thom (1952)

The Wardian Case:
a key vehicle for botanical globalisation

For centuries transporting plants across the oceans, such as coffee (from Yemen), tea (from China) and kina and rubber trees (from South America), was a rather hit and miss affair. Many plants died from salt pray, lack of light or lack of freshwater (which was scarce on sailing ships).

In 1829, the British physician and botanist Nathaniel Ward made an important discovery. Unable to grow delicate bog ferns in his London garden, he successfully grew them in a laboratory jar where they were free of smoke and pollution. He discovered that plants would grow there unattended for years without needing additional water or air. Only the correct amount of light needed to be determined. In 1832 he filled two glass containers with ferns and grasses and shipped them to Australia. After seven months at sea, they arrived in good health.

In 1842 Ward published the book *On the growth of plants in closely glazed cases*, that described his ability to keep plants in a bottle without watering for 18 years. The case was in fact a mini-glasshouse or terrarium, with glass sides and top, and was nearly airtight. If stored on deck or another place with sufficient daylight (detailed instructions were given on this), plants could survive for months without requiring additional water, while being protected from salt spray. The case was particularly suitable for pot plants.

The Wardian Case was a major innovation and a milestone in botanical globalisation. It had a considerable impact:

- Soon all of England adopted the cases and a national passion developed for exotic plants, particularly ferns and orchids, suited to growing in the sheltered environs of these cases. Many of the houseplants we know today were first collected and grown in Wardian cases.
- Whilst Darwin could not yet use the case on his voyage with the Beagle (1831–1836), Joseph Hooker was among the first plant explorers to use it, shipping many new and different specimens back to Britain during his voyage to the Antarctic (1839–1843).
- The British collector Robert Fortune used cases to ship 20,000 tea plants from Shanghai to the Assam region in India, where, to this day, much of the world's finest tea is produced.
- The Chinese banana *Musa cavendishii* was introduced to Fiji, where it started its global career.
- The Pará rubber tree from Brazil was shipped successfully to Malaya and Sri Lanka, thus establishing the British colonial rubber industry. Later, this proved to be an invaluable resource during two world wars and contributed significantly to the Allies' victory in World War II.

The Wardian Case, combined with the increasing speed of sailing ships, had a major impact on plant trade, gardening, agriculture and – indirectly - even wars.

Sources: Bailey (1928) and Chittenden (1951)

Replica of a Wardian Case in the Amsterdam Botanical Garden.

Aquarium trade as a vehicle for invaders: Caulerpa seaweed in the Mediterranean

Caulerpa taxifolia is a seafloor dwelling weed, originating in the Pacific Ocean. In 1980, the special characteristics of this seaweed were noticed at Stuttgart's zoo. The zoo had used the weed as an ornamental in its aquariums for many years. During this time it had been exposed to chemicals and UV light, and this is thought to have changed or activated certain genes. The plants did not wilt and grew at an exceptionally fast rate, even at low water temperatures. Word quickly spread, and soon aquariums from all over the world were requesting "cuttings".

One of these was the Oceanographic Museum in Monaco, in 1982. Two years later, *Caulerpa* was found to occupy a plot of about 1 m² in the Mediterranean Sea in front of the museum. Five years later, the species had expanded to cover an area of 1 hectare. From here it spread quickly. The species can now be found all over the Mediterranean Sea, using its enormous growth rate to compete away other seafloor-dwelling plants. It also benefits from being toxic to grazing animals. It is spreading through plant fragments (vegetative reproduction) that float away or are transferred by anchors, fishnets, ballast water or – again - the aquarium trade. It is thought that the species grows particularly fast in areas disturbed by contamination with urban wastewater (such as along the Côte d' Azur).

The negative aspects of the seaweed were recognised at an early stage, but only by few. The University of Nice has been particularly adamant about the need to control the species. Several methods have been tried (mechanical as well as biological, using algae-eating sea slugs), but so far with little success.

Meanwhile, *Caulerpa taxifolia* has invaded other parts of the world, including California, but there it could be eradicated before it could spread.

Source: Madl & Yip (2003)

Aquarium trade is a source of bio-invasions.

First account:
cholera in London 1830

Cholera invaded Western Europe in 1830. In 1854 the London physician John Snow did not yet see the bug but postulated the existence of *'Vibrio cholerae'*.

"...I suspected some contamination of the water of the much-frequented street-pump in Broad Street, near the end of Cambridge Street; but on examining the water, on the evening of the 3rd September, I found so little impurity in it of an organic nature, that I hesitated to come to a conclusion...

...On proceeding to the spot, I found that nearly all the deaths had taken place within a short distance of the pump. There were only ten deaths in houses situated decidedly nearer to another street pump. In five of these cases the families of the deceased persons informed me that they always sent to the pump in Broad Street, as they preferred the water to that of the pump which was nearer...

...We must conclude from this outbreak that the quantity of morbid matter which is sufficient to produce cholera is inconceivably small..."

Source: John Snow (1854)

Invasion success of pathogens
depends on multiple factors

The success of bio-invasions often depends on several factors. This is illustrated by the following examples of pathogenes.

The **bilharziasis** parasite, the trematode *Schistosoma*, uses freshwater snails as its intermediate host. In a new tropical area, the parasite has a chance of succeeding only if a specific snail has also settled there. Each species of *Schistosoma* depends on a different species of snail.

The **cholera** bacterium cannot develop into a major epidemic until it is infected with *two* different bacteriophages (Goudsmit 2001). Other factors also come into play. In the large Indian estuaries, the tongue of brackish water today penetrates much further inland due to the warming of the ocean water. This enables the bacterium to spread inland, exposing tens of million more people in these densely populated deltas to the disease.

Ships loaded with ballast water or changes in ocean currents like El Niño and La Niña may have caused the crustaceans carrying the dangerous cholera bacterium El Tor to be transported to the west coast of South America in the early 1990s (Wills 1996).

The **myxoma virus** introduced in Australia to control rabbits could spread only after mosquitoes were also infected with the virus.

The *Plasmodium* **species** causing **avian malaria** was present on Hawaii in introduced poultry kept by settlers around AD 1800, but it needed a vector to spread. This came with the introduction of the southern house mosquito *Culex quinquefasciatus* in the water barrels of a sailing ship in 1826 (ISSG undated). Many endemic birds went extinct (see Box 39).

The **yellow-fever virus** was able to spread through South and Central America only after two specific conditions had been met. First, a suitable species of *Aedes* mosquito had to arrive and occupy its own niche. Second, there had to be a considerable reservoir of people and animals (such as primates) infected with the virus. In the 1940s, these conditions were met: mosquito eggs and larvae arrived in water barrels onboard ships, in combination with the arrival of people carrying a mild form of yellow fever (Spielman & D'Antonio 2001).

Pathogens benefit from increasing ship speed

Throughout history, ships have spread plants, animals and pathogens across oceans. Up to around AD 1500, the Indian Ocean was the foremost ocean for global trade. After 1500, the Atlantic Ocean became more important. Over time, species benefited from the increasing speed of ships, allowing them to survive increasing distances. This is clearly illustrated by the distributional history of three human diseases: the plague, smallpox and cholera.

As mentioned in Box 5, the **plague bacterium** has caused three pandemics. Each of these was to some degree associated with ships. During the first pandemic the bacterium and its vectors (rats and fleas) were spread with ships across the Indian Ocean, the Western Pacific (as far as China) and the Mediterranean. During the second pandemic the pathogen and its vectors were spread with ships across the Mediterranean and Eastern Atlantic to Northern Europe. During the third pandemic the bacterium benefited from the global network of fast steamships developed in the 1870s.

The **smallpox virus** is related to some other poxviruses striking cows, monkeys and camels. Smallpox is a very old disease that may have developed in Egypt and surrounding areas. Egyptian mummies around 4000 years old show signs of smallpox; there are also reliable descriptions from India dating back to 300 BC. The virus invaded in Europe from Southern Asia. One epidemic in France in the 6th century is attributed to smallpox.

By AD 1507 the virus crossed the Atlantic for the first time, caused a severe epidemic in Hispaniola that killed most natives, but then died out itself. In 1518 a new out-break started among African slaves working in the silver mines of Hispaniola. It spread to Cuba and Puerto Rico, where it killed over half of the natives in 1519. In 1520 the virus jumped to the continent in a ship sent to Mexico from Cuba to find out how Cortés' troops were doing, after they had landed the previous year. The virus soon became Cortés' mightiest ally in conquering the Aztec capital Tenochtitlán. As early as 1568, the popula-tion of central Mexico had shrunk to about three million, i.e. to about one tenth of what had been there when Cortés landed. Measles, mumps and influenza also con-tributed to this decimation. Decay continued, though at a reduced rate, until 1620. Decimation also occurred in many other regions in the Americas, often with smallpox as the main killer

In the early years after 1492, the influx of viruses were limited by the slow speed of the ships crossing the Atlantic. Crewmen and slaves infected with smallpox had either died or recovered before arriving in Cuba or Mexico. In both cases the virus had died out before arrival. But with ship speed increasing, the survival opportunites for the virus increased as well. To prevent transatlantic infection, slaves would be transported to America only if they had become immune to smallpox – by being infected and surviving the disease. After 1700, however, immunity could also be acquired by inoculation with the human smallpox virus; and after 1798 by inoc-ulation with the cowpox virus. That reduced the influx of smallpox virus in the New World.

The **cholera bacterium** is another pathogen that ben-

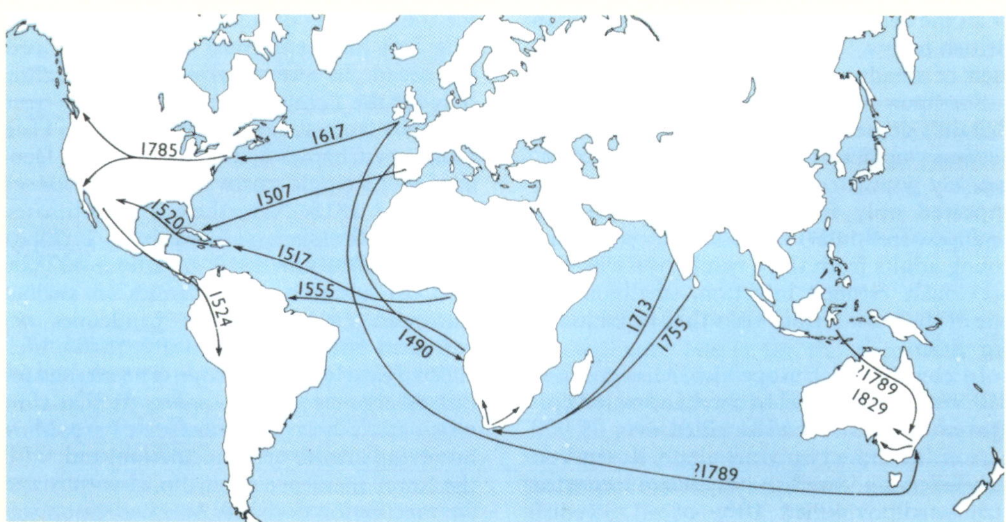

Figure 7.
The global career of the smallpox virus during the 15th-19th centuries (after Fenner *et al.* 1988).

Cholera in Paris (1840) by Daumier.

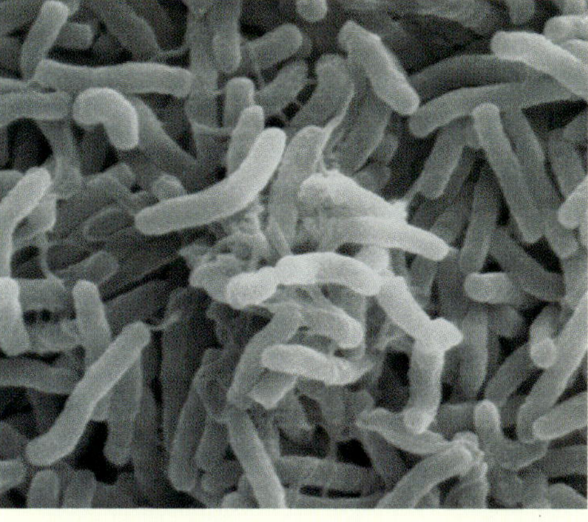

Cholera bacteria.

efited from the increasing speed of ships. Cholera was endemic in South Asia, particularly India, in ancient times. Although overseas trade between India and Portugal started shortly after Vasco da Gama's first visit in 1498, it took the cholera bacterium 330 years to reach Europe. Any patients onboard had either died or recovered before arriving in Europe. With the speed of sailing ships increasing, the bacterium could travel longer distances.

Between 1814 and 1975, there were seven cholera pandemics. The first (1817-1823) was confined to Asia reaching as far west as the Caucasus. In the early 1820s, ships carried the bacterium from India to Sri Lanka, Indonesia, China, Japan and Arabia.

The second pandemic (1826-37) reached Eastern Europe overland as far north as the Baltic. From there, it reached the British Isles by ship. Subsequently, ships with Irish emigrants helped the bacterium cross the Atlantic.

Two other factors helped the bacterium spread across the planet during the 19th century. Water came into use as ship ballast. And in the 1870s, a global network of fast steamships developed.

Of course many defensive measures were taken, including quarantine, but they did not prove very effective. A more effective response came when the relation with drinking water was discovered and slowly became accepted. During the second half of the 19th century, cholera gave a powerful impetus to many investments in water infrastructure, such as water pipes and sanitary water and sewage systems. In that respect, cholera became a history-maker.

However, the bacterium continued to strike, particularly after a new variety emerged, *Vibrio el tor*. It caused several regional epidemics before generating the seventh pandemic (1961-75), which hit Asia, North Africa, Spain and Portugal but not the Americas.

The South American cholera epidemic of 1991 – the first in America for over a century – may well have been caused by ballast water dumped by ships from Southern Asia (though an alternative hypothesis attributes the epidemic to sea currents such as El Niño).

This latter example perhaps illustrates a global change: after the Indian Ocean and the Atlantic Ocean, the Pacific Ocean is now becoming the foremost ocean for trade as well as biological invasions. And as a result of global warming, the fourth ocean may join in: the Arctic Ocean.

Sources for smallpox: Prescott (1901), Yanez (1950), McNeill (1976), Fenner (1988), Bol (1999a) and Childs Kohn (2001)
Sources for cholera: Pollitzer (1959), McNeill (1976), Wills (1996) and Childs Kohn (2001)

Car-tire transports transmitting dengue

The dengue virus causes a life-threatening disease, dengue fever, which leaves many patients seriously debilitated or dead. The primary vectors are mosquitoes of the genus *Aedes*, particularly *Ae. aegypti* and *Ae. albopictus.** From Africa, the disease has spread to Asia and Southern Europe, but the Americas have long remained free of dengue.

Aedes breeds in water; a can or jar is sufficient. This is why the "hygiene police" in stringent Singapore have the authority to inspect houses, balconies and gardens for places where *Aedes* can breed, and can impose high fines (up to US $50,000 and six months' imprisonment!) if such breeding sources are found. Special attention is given to scrap car tires filled with water, which *Aedes* has discovered to be an ideal nursery.

Countries where reuse of scrap tires by rethreading (remoulding) is prohibited for safety reasons, such as Japan and Taiwan, export scrap tires to countries where it is legal. These include Latin American countries. Brazil (ironically, the cradle of the rubber tree) is particularly hit hard by dengue; several hundreds Brazilians have already been killed. The US is also paying its debt. In the early 1980s, *Ae. albopictus* arrived in Houston, the scrap-tire capital of the world. Since then, dengue has been diagnosed in the US.

In the future, globalisation and climate change may cause these insect-borne diseases to appear in Northwest Europe as well (see Box 86).

* *Aedes aegypti* can transmit at least 17 viruses, including those causing various forms of encephalitis (like Japanese encephalitis) and yellow fever.

Source: Spielman & D'Antonio (2001)

Wars and pilgrimages as vehicles for infectious diseases

The major drivers of biological globalisation in the past have been trade, war and pilgrimage. In the last century, another major driving force has come to the forefront: mass tourism.

Examples of bio-invasions caused by **wars**:
- For thousands of years, Asian steppe peoples fought wars of conquest, even into Europe and India. In the 13th century AD, the Mongol rulers Genghis Khan and Kublai Khan ruled over the largest empire ever seen. Mongol invasions spread several diseases.
- Crusaders returning home helped spreading infectious diseases, including the plague in 1347.*
- Soon after its presumed introduction from America in Europe, syphilis quickly spread across Europe, partially as a result of war. Mercenaries who had participated in the French siege of Naples in 1494/95 spread the disease across Europe.
- During World War I, numerous Americans, Australians, New-Zealanders and South-Africans fought in Europe, among others. Their return home accelerated the Spanish flu pandemic of 1917-1919, which took the lives of at least 40 million people.

Examples of bio-invasions caused by **pilgrimages**:
- The Pope announced a Jubilee year in AD 1500, even though the plague had broken out in Rome and the surrounding area. One million Christians came together in Rome and many fell dead, often believing that death during pilgrimage would secure their place in heaven. The survivors spread the disease while returning home.
- From time immemorial, Hindu pilgrimages and times of festival have drawn crowds to the lower Ganges, where cholera was endemic.
- During most of the seven cholera pandemics (1817-1975), pilgrims visiting Mecca and Medina for the *hajj*, were infected and spread the disease to Islamic countries from Morocco to the Philippines while returning home. In 1905, a new strain of cholera was identified among pilgrims in El Tor, the quarantine station near Jeddah, the harbour for Mecca. The strain originated in Sulawesi, Indonesia. It was responsible for the seventh pandemic of cholera (1961-75) (Childs Kohn 2001).
- In 2001 and 2002, the *hajj* resulted in dozens of patients with meningococcal disease who had been infected with the rare W strain in Singapore and the Netherlands, among others. This strain is similar to serogroup C, which causes a lot of sepsis and mortality. In the Netherlands inoculation against this strain is currently taking place (Hijmans 2003).

* Incidentally, the crusaders also transferred some new agricultural crops to Europe, including the apricot.

The astounding transmission pathway of the foot-and-mouth disease virus in 2001

The transmission pathways of species are sometimes astoundingly complex. A clear example is the 2001 foot-and-mouth disease epizootic in the UK and the Netherlands.

The virus reached the Netherlands along the following complex route:

1. The first step of the transmission is the least clear. Meat carrying the virus was somehow exported to England, probably from the Far East, India, the Middle East or the eastern Mediterranean Region. Import may have been legal or illegal. Catering on airplanes has been suggested as a possible vehicle for the virus.
2. Near Aberdeen, the catering or restaurant refuse was delivered as swill to a pig farm.
3. The pigs became infected.
4. The pigs, in turn, infected cattle and sheep.
5. Contaminated sheep were exported to France, where they crossed the path of a shipment of calves to the Netherlands.
6. The calves were infected.
7. The calves, once arrived in the Netherlands, infected Dutch cattle.

The result was a traumatic and costly epidemic (see Appendix 3).

This case illustrates how complex the spread of a disease can be. Three livestock species (pigs, sheep and cattle) and three means of transport (airplane, truck and ship) were involved in the transmission.

The case also illustrates the risk of the globalising trade of living animals and meat, even when cooled or frozen. For example, in cooled bone marrow, the FMD virus can survive at least 5 months.

Finally, the case illustrates the risk of feeding animals with (inadequately treated) leftover food, particularly if that has come from far away.

The FAO recently expressed concern over the risks of livestock globalisation for several reasons including the spread of diseases and the damage outbreaks can cause for trade.

Sources:
DEFRA (2002),
http://www.agreport.com/open/139750.phtml

Culling of cows in 2001 in the Netherlands to eradicate the foot-and-mouth disease virus, which had invaded from Asia via Great Britain.

Early account: plant and animal invasions in New Zealand 1836

In New Zealand the rise of European species had already taken off by the time of Charles Darwin's visit, described in *The Voyage of the Beagle* (1836). That was just 35 years after the European colonization had started.

At first, Darwin writes with great satisfaction of the European flowers, vegetables, fruits, oaks, pigs and chicken he had encountered there. But then he notes that the brown rat had completely displaced the native rat (which we now know had itself been introduced by the Polynesians) within just two years. Finally, he seems embarrassed and becomes angry:

"In many places I noticed several sorts of weeds, which, like the rats, I was forced to own as countrymen. A leek has overrun whole districts, and will prove very troublesome, but it was imported as a favour by a French vessel. The common dock is also widely disseminated, and will, I fear, for ever remain a proof of the rascality of an Englishman, who sold the seeds for those of the tobacco plant."

Later, in *On the Origin of Species* (1859) Darwin reports:

"The endemic productions of New Zealand (..) are now rapidly yielding before the advancing legions of plants and animals introduced from Europe."

Patterns of bio-invasions

In bio-invasions certain patterns can be detected. We will consider the following questions:

- How do bio-invasions proceed over time?
- Which species are invasive?
- Which species are vulnerable?
- Which bioregions/ecosystems provide invasive species?
- Which bioregions/ecosystems are susceptible to bio-invasions?
- Can we, and should we control bio-invasions?

How do bio-invasions proceed over time?

The spreading of species over the planet often occurs quite slowly; sometimes it happens very fast, particularly where air traffic is involved. The quickest are viruses spreading from humans to humans through airborne transmission (droplet infections). Today such a virus can spread to every part of world within a few days. In most organisms, however, spreading is a much slower process.

As mentioned earlier, not every settlement leads to an invasion. Often a repeated influx of many individuals is needed before a species actually becomes established and spreads. For instance, the introduction of the myxoma virus to combat rabbits in Europe and Australia was successful only after repeated attempts. And the introduction of the sparrow in America succeeded only after several attempts: 60 individuals were placed in Central Park, New York City, in 1890 and another 40 the following year (Williamson 1997).[22]

After settlement, a species often starts reproducing at an exponential rate. That can happen slowly or rapidly, dependent upon the characteristics of the species itself (particularly the intrinsic rate of natural increase and the dispersion rate)[23] and upon the biotic and abiotic environment. A key factor in the biotic environment is the presence or absence of competitors, predators, parasites and pathogens.[24]

After this initial stage, we can observe the following patterns:

1. Many newcomers *become extinct* within a few generations. Apparently it is not sufficiently able to deal with the biotic or abiotic conditions, for example an irregularly occurring harsh winter or lengthy drought. In the case of a parasite or pathogen, the cause might be that the host species has become immune; and in the case of a host species, the cause might be that the species is attacked by indigenous pathogenic organisms, parasites or predators.

2. Sometimes there is a lag phase during which the species manages to sustain its numbers but does not spread in a significant way. This phase may last up to 100 years. For example, the Nile perch *Lates nilotica* was introduced in Lake Victoria in 1954, but remained little noticed until it exploded in the 1980s. The pinewood nematode *Bursaphelenchus xylophilus,* introduced from America, took even longer in Japan. Discovered in early 1900, it took 80 years before expanding explosively, at a cost of millions of pine trees (Bright 1998). Systematic analysis of fish, bird and mammal introductions in North America and Europe revealed that fish became settled on an average of 13 years after introduction, and began to spread around the same year. For birds, these figures are 52 and 60 years, respectively, and for mammals 14 and 62 years (Jeschke & Strayer 2005).

Multiple causes can be responsible for such time lags:

- Crossbreeding or mutation is required before the species is able to spread.
- Before the species is able to spread, it must "wait" for exceptionally favourable weather conditions (Hengeveld 2001).
- Among plants, the species may have to wait for the arrival of a crucial partner, such as a mycorrhiza, a seed-distributing animal or a pollinator. For instance, white clover *Trifolium repens,* which today dominates much of New Zealand's grasslands, was not very successful after its introduction in New Zealand. It was unable to reproduce, lacking a suitable pollinator. Its career started in 1838 after the honeybee was introduced (Crosby 1986).[25]

- Among specialist animals, the species may be able to spread only if an indispensable biotic or abiotic resource becomes available, for example by habitat transformation.
- Among pathogens and parasites, the species may have to first establish itself in a suitable host or intermediary host. Invasions are often determined by several factors (see Box 18). For the diphtheria bacterium, for example, it is not sufficient to be spread; in order to cause an epidemic, it must first be infected by a bacteriophage. The cholera bacterium must even be infected by two different bacteriophages before it can cause an epidemic (Goudsmit 2001).

Whatever the case may be, it can be safely assumed that there are dormant exotic species in many areas that will one day explode. But *which* species, we do not know.

3   In some cases a species will become dominant and *maintain its dominance*, leaving little or no space for local competitors. This phenomenon is known as "swamping". The Nile perch, for instance, became the dominant species in Lake Victoria, wiping out hundreds of endemic species (Goldschmidt 1994).[26] For other examples, see Box 29.

4   An explosion is often followed by a decline (a *boom-and-bust pattern*), followed in turn by stabilisation at a lower level or – rarely – by extinction. Examples from the Netherlands are the waterweed *Elodea canadensis* and the collared dove *Streptopelia decaocto*. Possible explanations for such a relapse are:
- Food or another resource becomes scarce.
- Infection pressure increases due to higher densities.
- Native parasites, pathogens or predators attack the new host.
- The invader is followed by a parasite, pathogen or predator from its native area. Sometimes this happens spontaneously, more often it is deliberately or accidentally facilitated by man (see Box 30).

5   In many cases, however, invasive species manage to maintain *higher densities than they did in their native area*. Often this is because they are released from (some of) their parasites, pathogens and predators (see Box 28). Another mechanism is biochemical warfare. Recent research has shown that invasive plants in North America contain relatively many phytochemicals that are toxic to herbivores, fungi, bacteria or other plants, and that are found neither (or rarely) in native plant species nor in

The Nile perch started swamping soon after its introduction in Lake Victoria.

non-invasive exotic species. For instance, root exudates of some invasive plants contain novel chemicals that are highly toxic to native plants (Cappuccino & Arnason 2005). One example is the hedge garlic *Alliaria petiolata* that kills fungi, including mycorrhiza fungi, in the soil. This gave the species a competitive advantage after it settled in Canada since – unlike the plant's competitors – it does not require the fungi (Van Strien 2004). Whilst the species rarely swamps in its native Europe, it now does so in many Canadian forests.

6   Some invasive species may even cause *radical changes in the biocoenosis, ecosystem and landscape, and possibly even the climate* (see Boxes 29 and 42). Such a species can be called a transforming species.

7   Some species spread *together with closely related invaders*, with which they may interbreed (Reumer 2005). Examples of such twin invaders into Europe are:

- The Japanese *Fallopia japonicus* and the giant knotweed *F. sachalinensis* from Northeast Asia.

- The North American pondweed species *Elodea canadensis* and *E. nuttalli.*
- The gallant soldier *Galinsoga parviflora* and the shaggy soldier *G. ciliata* from tropical America.
- The Asiatic clams *Corbicula fluminalis* and *C. fluminea.*
- The zebra mussel *Dreissena polymorpha* and the closely related quagga mussel *D. rostriformis bugensis*, both from Eastern Europe. These twin species are also invading North America.

Such twin invasions can be explained because the twin species occupy similar but not entirely identical niches

Changes following invasion

After having invaded, the species will often undergo changes, first phenotypic, then genetic changes.

1   In some cases *phenotypic changes in body size* occur soon after settlement. For instance, the species may become bigger as a result of losing some of its enemies (and thereby may cause more damage). However, it may also become smaller after it has reached high densities, due

The collared dove followed a boom-and-bust pattern of invasion in Europe, although the bust was quite limited.

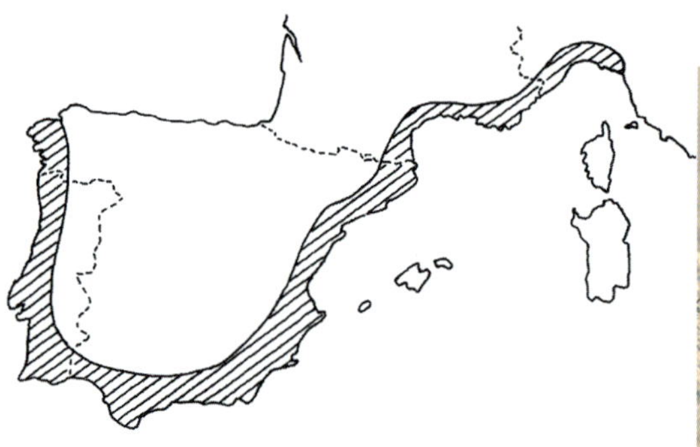

The Argentine ant forms super-colonies in invaded areas such as California and Southern Europe (From: Reumer 2005).

Argentine ants attacking a bigger ant.

to increasing pathogen or parasite pressure or to intra-specific competition. Such "stunting" has, for example, been observed in the pumpkinseed sunfish in the Netherlands (see Box 78).

2   The *loss of genetic variation* - an unavoidable side effect of any colonization – has the obvious disadvantage of making the species less flexible in adapting to its new environment or changes therein. But it may also have advantages. For instance, the smaller the variation, the less pathogens and parasites the species will carry. Other advantages occur, too. One surprising and spectacular example is the Argentine ant *Linepithema humile*. In Argentina, colonies of this ant species are kept apart by the innate aggression of the species against individuals of another colony. When the ant settled in California and Europe, this characteristic was partially lost. The result was the appearance of super-colonies that stretch hundreds of kilometres (in Europe even from Portugal to Italy), causing massive damage, even at the ecosystem level (Tsutsui *et al.* 2000; for other interpretations see Schwartz 2004). Research has started using pheromones to stimulate mutual aggression (Wassink 2006).

3   *Genetic changes* will occur through *natural selection*[27], both in the invading species and in indigenous species with which the invader strongly interacts:

- The invader adapts to its new biotic or abiotic environment and may, for example, begin to feed on new plant, prey or host species.
- Conversely, native prey species (or competitors or hosts) may adapt to the invader (e.g., by developing immunity or shifting their niche),

allowing them to partially recover.

- Similarly, native pathogens, parasites and predators can adapt to the invader, enabling them to attack with greater success. This may result in a co-evolutionary "arms race" in which the pathogen or parasite often has the advantage of having a higher rate of generation turnover than its host.
- If the colonizing species is itself a (monophagous) pathogen or parasite, it can become *less* virulent. The most virulent varieties may go extinct because they do not allow their host enough time to spread the pathogen or parasite.[28]
- In the long term, mutation, recombination and selection can lead to the creation of new species. But this may well take more than 100,000 years (depending on the speed of generation change), and has so far not been reported in anthropogenic invasions (Williamson 1997).

4   Occasionally, an invasive species will *crossbreed with an indigenous species*. This can have a variety of consequences. For instance the genome of the native species may be displaced, as has been found in some fish and duck species. A current example is the ruddy duck *Oxyura jamaicensis*, an American species that crossbreeds with the rare, endangered white-headed duck *O. leucocephala* in Southern Europe, threatening its survival as a distinct species. It is also possible, however, that a new (sub) species originates. This is what happened when an American species of spartina grass crossbred with a European species in the British Isles, creating an extremely invasive hybrid that even spread globally (see Box 29). The same thing happened in Brazil when the African subspecies

of the honeybee was crossbred into the population of the previously introduced European subspecies of honeybee (see Box 55). The result was a monster: a highly aggressive and invasive new subspecies that spread north as far as the southern US.

5   In some cases an invasive species *paves the way for other invaders*. For example, the zebra mussel is turning bodies of water in America into suitable environments for invasive water plants (see Box 57); and in the Netherlands the Japanese oyster provides excellent substrate for an invasive colonial tunicate, *Didemnum cf. vexillum,* as well as for other species.

Which species are invasive?

Species and species groups differ greatly in their invasive ability. Elton (1958) for instance noticed that no birds from Eurasia had successfully settled in American forests, whereas many Eurasian insects had done so, including some harmfull species. Globally, beetles are a relatively successful group (Williamson 1997).

Ecologists have made several attempts to predict how invasive species are. Such research however has yielded little so far. Predictive theories are still at an early stage.

Taxonomy also offers little to go by. Invasive species appear to occur in a wide range of taxonomic groups. An invasive species is often the only one in its family, indicating that small taxonomic differences can make dramatic ecological differences. This is a fact well known to experts in pathogens of plants, animals and humans. Of two varieties of the same species, not distinguishable by classic morphological methods, one can be a serious pest or pathogen while the other is harmless. Experimental research has shown that transplanting a single gene from a pathogenic *Yersinia* bacterium into a harmless *Escherichia coli* bacterium may make the latter into an invader of cells in culture. Likewise, the collared dove *Streptopelia decaocto* has invaded Europe, whereas closely related species, with which it can easily interbreed in captivity, did not (Williamson 1997).

Vermeij (1996) distinguishes between three stages of invasion: arrival, settlement and integration (fitting in to the new biocoenosis). A plausible hypothesis is that in order to be successful an invasive species must be "good" at *every* one of these stages; this hypothesis however is as yet untested. "Flexibility" is said to be another trump card, but there is also little evidence for this at present. Not even the matching of species and climate will always be an accurate predictor. Once arrived in a new environment, some species spread well beyond their normal climatic limits. This may have to do with differences in the biotic environment.

According to Williamson (1997) there are only three reliable predictors:

- The species fits into one of the present habitats.
- The species had already been found to be invasive in another region.
- The species is able to exert significant propagule pressure.

Ricciardi and Rasmussen (1998) added two characteristics:

- The species has a good chance of "hitching a ride" by a means of transport.[29]
- The species is able to withstand the stresses of transportation. For example, a species with a broad salt tolerance is able to survive the (partial) replacement of ballast water.

They argue that the most consistent feature of invasive species is commensalism with humans.

Based on these criteria they made some predictions. First they prepared a list of all possibly invasive European species in North America's Great Lakes area. By examining the effects that invasions of these species had caused in other regions, they estimated the risks of system disruption that would occur if one of the species invaded the Great Lakes area.

The Global Invasive Species Database[30] uses the first three characteristics mentioned above to construct a habitat-matching model.

It has also been postulated that species adapted to high-resource environments are relatively successful invaders because they benefit more from escaping their natural enemies than species in low-resource environments (Blumenthal 2005).

Furthermore, common sense suggests that invasive species do *not* have to be able to spread over great distances, at least not where anthropogenic invasions are concerned. Furthermore, other things being equal, *small* organisms (or organisms with small propagules) are at an advantage over larger organisms, at least during the early phases of the invasion process: transport and settlement. Reasons for this are:

- Their *small size* gives them more opportunities to hitch a ride as "stowaways"; and being harder to notice, they are harder to control.
- They are often *numerous*, providing them with more opportunities to "hitch a ride". They also have a greater chance of breeding with a partner following arrival and less chance of being completely wiped out by predators, parasites, pathogens, competitors or humans. In addition, they have more genetic variation and thereby more opportunities to adapt.

- They often *reproduce more rapidly* and thereby need less time to develop a viable population.[31]

However, this does not necessarily hold for the subsequent phase of maintaining, reproducing and spreading. For instance, once arrived, a small plant species may have a competitive disadvantage. Successful exotic plant species are often large. Perhaps a large plant producing small seeds or spores may make a successful combination.

Invasive species have often taken biologists by surprise. A species that takes a modest place in its native area may swamp in a new environment. Box 29 gives a few examples.

Which species are vulnerable?

A little more can be said with certainty about species that are *vulnerable* to bio-invasions. Often, these are species with a small genetic base or gene pool, which makes it more difficult to adapt to an invader. Such species are often found in small, remote bioregions, such as islands and lakes.

All species become more vulnerable to invaders in periods when their populations are weakened, e.g. by parasites, food shortage or extreme weather conditions.

Invasive species can themselves be vulnerable, too. Because they often have fewer natural enemies in their new area, they are able to expand relatively fast. But of course they remain vulnerable if one or several of their natural enemies also manage to cross over into the new territory, with or without the support of humans.[32] Over time, they also risk losing their defences from pathogens and parasites from their native region.

An example of a wild species is the ctenophore *Mnemiopsis* in the Black Sea, which first exploded but then was decimated by another ctenophore that subsequently invaded (see Box 30). A classical example of a vulnerable introduced crop is the potato, which after a successful career was devastated in Ireland in AD 1847 after the fungus (or rather: oomycote) *Phytophthora infestans* managed to hitch a ride across the Atlantic (see Box 45).

If the natural enemy itself has natural enemies, it also may be vulnerable. In some cases researchers have made clever and thankful use of that, mainly in agriculture and horticulture.

In some cases, an indigenous species can benefit from an invader. For example, indigenous scale insects on trees and other plants in the US benefit from the invasive Argentine ant. This invader protects them from their predators, enabling them to increase their numbers. For the plant, however, this obviously is a disadvantage. Likewise, the endemic Kea parrot *Nestor notabilis* of New Zealand has discovered sheep carcasses, and occasional-ly even living sheep, as a source of meat. But of course, the bird since then risks the wrath of sheep farmers.

Which bioregions and ecosystems provide invasive species?

An interchange of species between bioregions is more likely to happen between two bioregions with a similar climate than between two bioregions with a completely different climate. Most species are simply poorly equipped to survive in a different climate.

There are, however, some exceptions. For instance, the chufa flat-sedge *Cyperus esculentus* (also known as tiger nut or yellow nutsedge) stems from the tropics but has become one of the most dreaded weeds in many parts of the world. In the Netherlands, it is a serious pest in the cultivation of maize and flower bulbs, and is actively combated there. The rose-ringed (or ring-necked) parakeet *Psittacula krameri* and the Egyptian goose *Alopochen aegyptiacus* – both introduced from the subtropics, held in captivity, and then released or escaped into the wild – have managed to thrive and multiply in Britain as well as the Netherlands.

A special case is microclimates in heated buildings, greenhouses and surface water, which offer opportunities to subtropical or tropical species. For example, Dutch greenhouses have become home to such species as the Med fly and Californian thrips. In buildings we find species such as cockroaches and the longhorn beetle; and near cooling water outlets from power stations and factories, tropical fish such as guppies can often be found. These fish were in most cases dumped by aquarium owners. Such settlements do not always remain local, though. For instance, trade may spread plant diseases from greenhouses to other countries. This can even lead to an import ban.

Species interchange may take place in both directions, but is rarely symmetric. Islands, for example, sometimes deliver invasive species to neighbouring islands[33], but rarely to continents. Among the few cases are the Black Sigatoka virus of the banana plant that originated in the Fiji Islands[34]; and the El Tor variety of the cholera bacterium that originated in Sulawesi, Indonesia (Childs Kohn 2001). Of course, the bacterium originated not on land but in the (brackish) waters surrounding Sulawesi, but we can consider these waters as an "island" as well.

Even small continents provide more species than islands. For instance, trees belonging to the genera *Eucalyptus, Acacia* and *Melaleuca* have been introduced from Australia to other regions, where they have become serious pests.

Asymmetry is also often observed where seas, rivers or lakes were connected. For instance, following the construction of the Suez Canal, more species have entered the Mediterranean Sea from the Red Sea than vice versa. There are several explanations for this:

- The Red Sea is connected to the species-rich Indian and Pacific Oceans, while the Mediterranean is in connection with the Atlantic Ocean, which is less rich in species.
- The Red Sea is more saline than the Mediterranean, making it easier for species coming from the Red Sea to pass the barrier formed by two hyper-saline lakes through which the canal passes.
- There are more fishermen from the Mediterranean who fish in the Red Sea than vice versa. The fishermen clean their nets in their homeports – a recipe for introductions.

Following the completion of the Rhine-Main-Danube canal in 1992, more freshwater species have migrated from the Black Sea area to the Rhine than vice versa. This is mainly because the dominant flow of water in the canal is in the direction of the Rhine.

Generally speaking, the following patterns can be identified[35]:

1   *Species-rich regions* provide more invasive species than species-poor regions. Often these species-rich regions are larger in size; the enormous Eurasian continent for instance has provided more species to America and Australia than vice versa. The most obvious explanation for this is that species-rich regions simply have more species to provide. Furthermore, the more species there are, the higher the selection pressure, and therefore the greater the competitive force of the species (Darwin 1859).

2   The Eurasian continent, being *oriented East-West*, has produced more invasive pathogens than the American continent, which is oriented North-South. Diamond (1999) provided the following explanation for this: in Eurasia a greater amount of migration has been possible because species were and are able to migrate over enormous distances while staying within a single climate belt. This raises infection pressure and induces selection for more natural resistance against more diseases. The combination of diseases and resistance became a deadly weapon when Europeans set out to conquer other bioregions including the Americas and many islands.

3   A region that *exports ecosystems* also exports associated species. This applies in particular to the export of agrarian and urban systems from Europe to North America, Argentina, Australia and South Africa. Europe thus became an "ecological imperialist" (Crosby 2000).[36] European agricultural systems are characterised by intensive grazing and/or a radical disturbance of the soil. Indigenous vegetations were often unable to survive such intensive use and declined rapidly, offering vacant niches to numerous grazing-tolerant invasive European species.[37]

4   For pathogens, the rule is that the *native region of the host species* contains and provides more pathogens of that host species than the region where the species is introduced. For example, most potato diseases have developed in America, wheat diseases usually stem from Eurasia and chicken diseases generally come from East and Southeast Asia.

5   Also for pathogens, *the higher the density and population size of the host species* (i.e., the more crowding), the higher the infection pressure and the higher the chance that *new* pathogens emerge.[38] In humans and livestock, this applies in particular to Southeast and South Asia. McNeill (1998) postulated that from 500 BC to AD 1200 the "civilised disease pools" of the different city networks in Eurasia underwent confluence to form one huge Eurasian disease pool.

6   For zoonoses it seems that *regions where large numbers of humans and animals live in close proximity to one another* produce more *new* pathogens than regions where the density of livestock and/or humans is low (Diamond 1999).[39] Such combined high densities are particularly common in Southeast Asia (as well as in the Netherlands, although in that country the numbers are much smaller and there is much less crowding). The risk increases where people are in close contact with each other near houses and in marketplaces (Bol 2003a, 2003b). It becomes even higher if different species of livestock are involved – such as pigs and poultry in China – offering more opportunities for genetic recombination.

These patterns combined help explain what to Darwin (1836) came as a shock:

"Wherever the European has trod, death seems to pursue the aboriginal. We may look to the wide extent of the Americas, Polynesia, the Cape of Good Hope, and Australia, and we find the same result."

Which bioregions and ecosystems are susceptible to invasions?

Vermeij (1991) studied 12 biotic interchanges that occurred in history, including those between the Pacific and Arctic Oceans after the formation of the Bering Strait, and between the Red Sea and Mediterranean Sea after the construction of the Suez Canal. His main conclusion was that relatively impoverished biotas are more susceptible to invasion, and that impoverishment often comes from extinctions *before* the exchange, creating empty niches. However, Williamson (1997) accurately remarks that this asymmetry might be merely a matter of chance: species-rich biotas *a priori* have better opportunities to produce invasive species.

From all this, it should *not* be presumed that only species-poor and small bioregions such as islands and the Australian continent are susceptible to bio-invasions. Bio-invasions can also occur in species-rich bioregions and on large continents. As Darwin remarked in 1859:

> "Probably no region is as yet fully stocked, for at the Cape of Good Hope, where more species of plants are crowded together than in any other quarter of the world, some foreign plants have become naturalised without causing, as far as we know, the extinction of any natives."

However, some of the numerous additional invasions that have taken place since Darwin's times have become a threat to the biodiversity in the Cape Province.

North America by now counts some 50,000 invasive species, including microbes (Pimentel *et al.* 2005). Apparently even the largest continent, Eurasia, is not yet fully saturated with species. This is evident from, for example, the invasion of the muskrat, which was repeatedly introduced from America to Europe for fur production and ultimately escaped into the wild near Prague in 1905. The species has since spread over a massive area.

We can identify the following patterns[40]:

1   Tropical continental regions appear to be less susceptible to invasions than subtropical and temperate continental regions. Although they have more species generally, they have fewer exotic species. This could be explained by assuming that fewer species have been introduced, but that does not seem very likely. For example, we can safely assume that Europeans have attempted to introduce many species in central Mexico. A more plausible factor is that biotic pressure from pathogens, parasites, herbivores and predators is relatively high in the tropics, making it difficult for any exotic species to establish. It may

be no coincidence that most exotic species in the tropics live in urban environments, where biotic pressure is less high (Sax 2001).

Two statistical factors can help explain the difference too. First, since the tropics are relatively rich in species, the pool of *exotic* species is by definition relatively small. Second, roughly speaking there are 3 major tropical bioregions in the world (in South and America, Africa and Asia) as opposed to 6 subtropical and temperate regions, each with a different evolutionary history. Hence, a given tropical bioregion can obtain species from 2 other pools, whereas a given subtropical or temperate region can obtain species from 5 other pools.

2   *Small and remote* bioregions such as Australia, islands and lakes are particularly susceptible. Explanatory factors include:

- *Chance*: the smaller the surface area of a bioregion and the more remote it is from large continents (species reservoirs), the fewer species the region accommodates[41] and, hence, the larger the chance that a colonizing species is new.

- *Evolution: the fewer species, the less competition and the less infection pressure*. Over time, a species' competitive abilities and its natural defences will weaken. For example, a recent study of the natural defences against parasites of an endemic ground finch on the Galapagos Islands showed that the smaller the island, the weaker the bird's natural resistance (*Der Spiegel* 2004).

- *Gene pool:* island populations often descend from a small group of founders; such populations have a small genetic base and hence limited adaptability. Native Americans also descend from a small group of founders and still have limited genetic variation, including variation in blood groups (Crosby 1972). That must have been one of the reasons for their high vulnerability to the infectious diseases brought by Europeans and their African slaves.

3   *Oceanic islands* are more susceptible than continental islands (Elton 1958) because they were never connected to a continent, so fewer terrestrial and freshwater species were able to colonize them. They have fewer species and more empty niches than might be expected based on the size of their surface area. The same probably applies to *lakes* that have not been connected to another lake for a prolonged period of time.

4   *Young ecosystems* are particularly susceptible because they are only partially colonized. Wallace reported already in 1880 that in Britain as well as New Zealand, species (both native and exotic) that do not naturally occur in a particular area, might suddenly appear along recently constructed railway lines or burned-off plots of land.[42] Cities and greenhouses, too, can be considered young ecosystems, with an increased chance of invasions.

On a geological time scale we can see that when the last ice age ended and the ice retreated, young ecosystems with plenty of vacant niches were formed. Most likely, the colonization of these niches is still incomplete in many areas. This also holds for marine ecosystems. When after an ice age the sea level rose again (by 135m after the last ice age), and land became submerged, many empty niches had to be replenished. In some parts of the world, that could happen quickly from species reservoirs. In other parts of the world, the process had to start from scratch, leaving plenty of room for exotic species. This is what happened in the Caribbean Sea.[43]

The North Sea is even younger (8000 years). Consequently the system is not yet completely balanced, leaving many niches for further introductions. Additional opportunities are created by the dynamic, partially even estuarine character of the area. A third factor is that the water temperature in the North Sea has risen quite slowly since the last ice age, and is still rising; this means that the ecological conditions are still changing. These factors have prevented "saturation" of the biota with species (Hofstede & Wolff 2002, cited in Tien & Dankers 2004).

5   Many authors emphasize the key role of *disturbance* (both natural and human-induced) in the susceptibility of ecosystems. Numerous examples have been given, starting with Wallace's observations on plants mentioned above. A more recent example is the deliberate introduction of the red king crab *Paralithodes camtschaticus* in the Barents Sea following over-fishing of native species (Reumer 2005). Perhaps the success of the muskrat in Europe may at least partially be explained by the previous extermination of predators like the European mink *Mustela nutreola*. Plant research by Lake & Leishman (2004) in Australia suggest-

ed that disturbance is a *sine qua non* for invasions.

6   *Resource abundance* is another key factor for the chances of invasions. Lake & Leishman (2004) showed that systems rich in nutrients and water are much more susceptible to plant invasions and have fewer native species than systems poor in those resources, even if the latter systems have been disturbed.

Davis *et al.* (2000) have put forth a somewhat different hypothesis that offers testable predictions. They suggest that *fluctuations* in resource availability can be the key factor for an ecosystem's susceptibility to invasions. The intensity of competition is inversely proportional to the amount of unused resources. This implies that every factor causing an increase in availability also leads to an increase in susceptibility. Changes in resource availability constantly occur in all kinds of systems. This happens in two different ways: either usage by the local populations decreases (for one reason or another); or the resource increases more rapidly than can be utilised by the local populations (again for one reason or another). A colonizing species can expand only when enough germs of the species are present at the moment of increased availability. Susceptibility to invasions thus fluctuates over time.

7   Elton postulated in 1958 that *species-poor* ecosystems are *as such* more susceptible to bio-invasions than species-rich ecosystems are. However, little empirical evidence has been found to support this hypothesis (Williamson 1997). For instance, many extreme ecosystems, including Arctic systems, are poor in species but rarely invaded. The diversity/stability hypothesis, though popular in the 1950s and 1960s, currently has few supporters among ecologists (Simberloff 2000).

Can we, and should we, control bio-invasions?

There has been much debate on the question whether we *can*, and if so, *should* prevent, reverse or control bio-invasions. There are no simple answers to these questions.

Should we try and prevent invasions? There is at least one category for which the answer is "no": invasions driven by climate change. Here it may be wise to allow species to adapt to the new physical situation. Stopping them may on the one hand trap the species in an unsuitable climate, endangering their survival, and on the other hand create incomplete and unstable biotas, poorly matched to the climate. Furthermore, the risk that such species will explode is limited, since they will often be joined by their natural enemies. In those cases where distributional barriers prevent spreading, we might even consider *helping* species spread. This can be done by translocating some species or by creating ecological corridors for all species.

However, in the case of the breakdown of a natural barrier, prevention may be preferred in order to limit mixing of biotas and invasion of pest species. But, these species, too, will sooner or later be joined by natural enemies checking their numbers.

Prevention may be even more desirable in the case of invasions resulting from trade and traffic. Often the species involved will be accompanied with few natural enemies, increasing the risk they will dominate and replace indigenous species. Furthermore, these species may come from distant and very different biotas that we may not like to see being mixed.

In which cases *can* we prevent invasions? Generally speaking, species approaching in a broad front are almost impossible to stop. This is often the case in invasions driven by climate change. Particularly hard to stop are airborne and arthropod borne pathogens. But if species need passing a bottleneck, such as a seaport or an airport, we have a realistic opportunity to prevent invasion. This is sometimes the case with the breakdown of a natural barrier, and often with international trade and traffic. Here we can create more or less effective distribution barriers, such as border checks and quarantine.

Another case where prevention is possible is when species shifting their range in response to climate change, are blocked by a continuous zone of unsuitable habitat. Then we can selectively translocate some (or all) species, or create an ecological corridor for all species.

There are several considerations in favour of the chances for prevention:

- A significant proportion of past harmful invasions resulted from intentional introductions, even repeated "pushing" of the species. Such mistakes can be avoided in future.
- Invasion risks are largely concentrated on specific high-risk flows of transportation, such as ballast water, animals and plants with soil attached; and of these, particularly flows between regions with a similar climate.
- Even "flash" invasions can sometimes be controlled. For instance, in 2003 a globally coordinated campaign against the new emerging disease SARS contained and eradicated the virus within a month.

Table 3 summarizes the argument. It seems clear that the focus in invasion prevention should not be on climate change, but on trade and traffic.

Here we have to make an exception. The assumption that those species invading in relation to climate change or the breakdown of a natural barrier will be joined by their natural enemies does not hold for viruses. They have few natural enemies, their main enemies being the antibodies and leucocytes of animals, which will not join the viruses while they travel north. Since these viruses are hard to stop and control, the best strategy may be to improve hygiene and to enhance the innate (general) and

Table 3. **Options for prevention in three categories of invasions**

Invasions related to...	Species needs passing a bottleneck	Natural enemies joining	Prevention advantageous in...	Prevention possible in...
Climate change	Rarely	Most or all	Few cases	Few cases
Breakdown of a natural barrier	Sometimes	Most or all	Few cases	Some cases
Trade and traffic	Often	Few or zero	Many cases	Many cases

acquired (specific) natural immunity of animals and humans, e.g. by vaccination.

If prevention proves unsuccessful, an invading species can, at least in theory, be halted either by eradication or by containment; and if prevention, eradication and containment are impossible or fail, the last option is to control the numbers of the species.

Meanwhile it should be clear that the prevailing strategy of nature conservation, the creation of nature reserves, does not offer much protection against bio-invasions. Particularly not if large numbers of people and goods are allowed to move between continents without proper border checks (see Box 34).

Prevention

Attempts to prevent invasions by way of traffic or transport can be very effective. Some species, like the rabbit in Australia, the sparrow in North America and certain trees in South Africa, were repeatedly "pushed" into the recipient ecosystem before achieving success. Such invasions are by definition easily avoidable.

Regarding species transferred accidentally: of course it is not feasible to check every single traveller or trade item on every thinkable invasive species. But as mentioned above, risks are fairly concentrated in particular flows. We are far from powerless to prevent such invasions.

It is possible to perform strict border checks of persons, cargo and means of transport, and to enforce quarantine if necessary. One problem here is that national borders rarely coincide with borders of bioregions. Consequently, it may be better to perform checks at departure than at arrival.

In addition, certain technical prevention measures are possible, such as:

- Placing barriers like discs and funnels on ship ropes to prevent rats from entering ships.
- Placing decontamination mats in airports to prevent the import of diseases that can be carried on shoe soles (in the Netherlands such mats were used against bird flu in 2003).
- Vaccination of humans and animals.
- Decontamination of plants, animals and wood (including pallets, packaging wood and ballast wood) using broad-spectrum biocides such as methyl bromide.
- Decontamination of ship's hulls using anti-fouling chemicals such as tar, tributyltin (TBT) or copper paint.

All of these methods carry a financial cost, and decontamination may also have environmental effects. All chemicals involved are strongly regulated or have even been banned.[44]

Eradication

Once an invasive species has managed to settle it can, in principle, be eradicated. This, however, must happen quickly; once the species has spread, eradication may be a mission impossible (e.g. the *Phytophthora* oomycote and the muskrat)[45] or a very costly mission (e.g. the foot-and-mouth disease virus).

A fair number of success stories have been reported (see Box 46). With few exceptions, these successes were achieved on small islands and on a small number of invasive species, namely rats, cat and rabbit. As anticipated, such eradications often lead to a recovery of the biodiversity, particularly among ground-breeding birds.

However, success stories from large islands and from continents are rare and mostly limited to species that had spread only locally. For instance, the invasive muskrat was successfully eradicated in Great Britain and Ireland in 1935 (Williamson 1997).

Among the rare success stories on continents, two - perhaps surprisingly - did focus on insects (see Box 54):

- The malaria mosquito *Anopheles gambiae*. Introduced from Africa in Brazil in 1929, it was eradicated in the 1940s by large-scale but targeted use of chemicals.
- The New-World screwworm fly *Cochliomyia hominivorax*. After it had crossed the Atlantic to form a beachhead in North Africa, it was quickly eradicated. The sterile male technique was applied.

Perhaps even more surprisingly, the one and only global success story focused on a virus: the smallpox virus. It was eradicated in 1977 after a 10-year programme of the WHO, mainly based on vaccination. This programme could succeed only because the virus has no reservoir in wild animals.

Many campaigns have failed, however, due to a lack of funds and legal regulations; problems with governments; delays; or even active opposition from local animal rights and conservation activists (Genovesi 2005). In one painful example, Italian animal rights activists prevented a local invasion of grey squirrels (native to America) being halted in its early phase. This allowed the species to begin spreading over the Eurasian continent and displace the indigenous European red squirrel, as it had already done earlier in Great Britain (Bergmans 2001).

Containment and control

Once a species has expanded, it may in some cases be possible to contain it, using natural or artificial dispersal barriers such as fences, cattle grids or ecological "fire-lanes", e.g. a ditch or forest. Such barriers will have to be

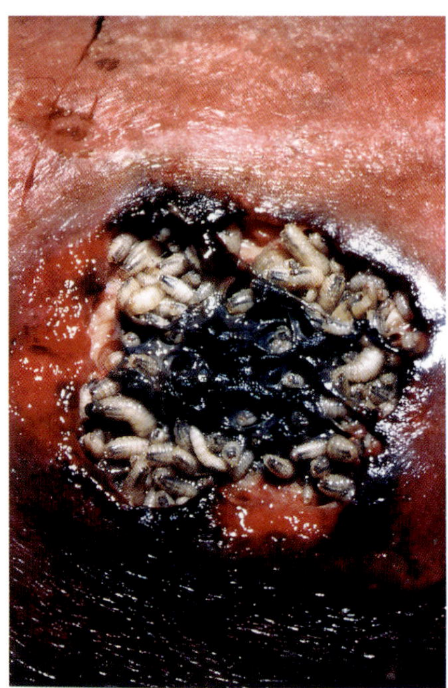

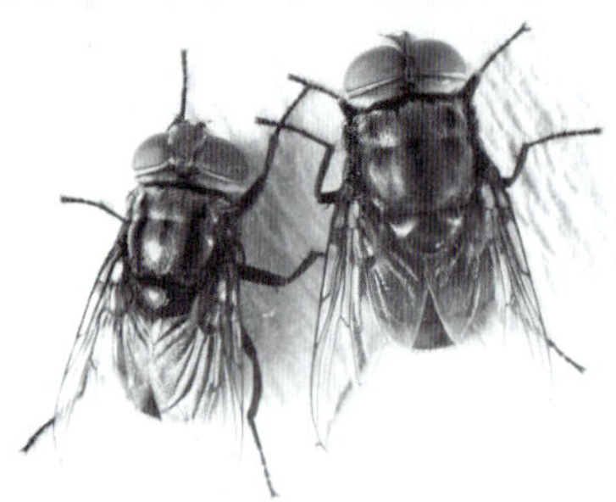

The New-World screwworm fly can cause horrible wounds in mammals including humans. After it formed a beachhead in North Africa, it was quickly eradicated (From: Clyti *et al.* 2003).

maintained indefinitely, though; and there is always the risk of some individuals escaping.

If containment is impossible or has failed, only a single option remains: control. This can be done using chemical, mechanical or biological means. Some examples:

- *Chemical control* of agricultural pests, water plants and growth on power station intakes. In the Netherlands, for instance, water pennywort *Hydrocotyle ranunculoides* is now combated chemically, using MCPA.
- Introduction of *natural enemies* of the species or of related species. This can be regarded as a planned counter-invasion. The best chances of finding a competent natural enemy are in the native region of the pest species or its host.[46] A classic success story is the introduction of an Australian ladybeetle in California for control of scale insects in citrus plantations (see Box 31). Africa, too, has seen successes. Of the 14 most important pests in agriculture and water, most have now been partially or completely brought under control using biological methods, including 7 that were controlled by introducing natural enemies (Bokonon-Ganta 2001). One of these successes was achieved in the 1980s, when a South American parasitoid wasp was introduced to control the invasive cassava mealybug (see Box 31).

Sometimes competent natural enemies are found outside the native range of the pest species or even in a different host species. For example, in the search for a pathogen for rabbit control in Australia, the myxoma virus was found in South America, not in the rabbit but in a related species. The virus proved successful, though its effectiveness declined over time. More recently, the RHD virus from China proved effective, too.[47]

On the other hand, many introductions of natural enemies have become failures or have had adverse side effects. This even happened in the case of some strictly monophagous enemies (see Box 32).

- *Other biological methods*, such as sterile males and pheromone-laced traps.
- *Mechanical methods*, such as the uprooting of plants or the catching of animals. In the Netherlands these methods have been used with varying success against the black cherry and muskrat.
- *Cultivation-specific methods*, such as the suspension of sprinkling. This method is being used with some success in the Netherlands to control the invasive bacterium *Ralstonia solanacearum*, which causes bacterial wilt in potatoes (Tönjes 2005).
- Control of infectious diseases of humans and

livestock by *hygiene, vaccines*[48], *antiviral drugs or antibiotics*. Since the 1980s, more and more effective antiviral drugs have become available to combat viruses. HIV is a good example.[49] However, effective drugs and vaccines against bird flu (subtype H5N1) are not yet available, partly because the virus frequently changes (antigenic shift).

- *Combinations of methods, for example of pheromones and toxic chemicals.* This combination has been applied successfully in the Netherlands to control the invasive tropical pharao ant.

All of these methods carry a cost in both money and labour, and many must be continued indefinitely. Most have other disadvantages as well, such as damage to the environment (chemicals), risks to public health (chemicals, antibiotics) or damage to native species (polyphagous natural enemies, side catches in traps). In theory, these costs and disadvantages are the least if the method used is the introduction of monophagous natural enemies. Such an introduction may carry a financial cost just once, and often causes no harm to other species, environmental quality, and public health.

A serious problem is that one cannot always predict for sure whether or not a species is monophagous, not even through testing. Furthermore, even truly mono-phagous enemies may cause damage through unexpected ecological interactions. In Australia, for instance, the gall wasp *Trichilogaster acaciaelongifoliae* was introduced from South Africa to control a species of *Acacia*. Although tests were performed before the species was introduced, it turned out that the gall wasp also attacked two other species, including a species of *Acacia* used for wood production. Possibly, the gall wasp underwent evolution following its introduction (Williamson 1997).

An even more striking example was recently described in North America (see Box 32). Two gall fly species introduced to control invasive spotted knapweed (native to Europe) did not effectively control the weed, thereby could become abundant, and attracted deer mice. These mice carry a hanta virus. This way the introduced flies stimulated the spreading of a zoonosis! The authors conclude that a natural enemy that is a candidate for introduction should not only be tested on being monophagous, but also on being effective.

If an introduced natural enemy becomes a pest itself, we may be left with permanent and irreversible damage. This may include damage to indigenous species, to the environment (by chemical control), and to the economy (losses, damage and control costs); and whilst a harmful chemical can be withdrawn, or substituted by a less harmful chemical, the alien pest species will likely be there to stay.

Introduced crops and livestock becoming pests

It is often assumed that introduced agricultural crops and seedlings do not invade non-agricultural land. This is true in many cases, since during domestication these plants were selected for the optimised environment of cropland. However, there are many exceptions too. Some examples:

- As early as 1836, Darwin mentioned the rampant growth of **cardoon *Cynara cardunculus*** (a predecessor of the prickly artichoke) in Argentina. Crosby (1972) speculates that this invasion could have been facilitated by overgrazing.
- Darwin also mentions swamping of **leek** in New Zealand.
- Of 466 species of **feed crops** introduced in Australia, 60 have become invasive.
- **Johnson grass *Sorghum halepense*** was imported as a crop from the Mediterranean area to the US; there, it has developed into a formidable weed (Pimentel *et al.* 2005).

- **Kudzu *Pueraria lobata*** was introduced in the US from Japan because of its edible roots. Nowadays it is a pest in the southeastern US where it has even become the plant covering the largest surface area (Pimentel *el al.* 2002).

Introduced trees, too, have often invaded surrounding areas (see Box 46).

The same goes for livestock introductions. Numerous examples are known from islands, where goats, sheep and pigs have caused much ecological damage. But continents have seen similar events. Cattle introduced in the Americas went wild in many grassland areas and their numbers exploded, leading to overgrazing. Numbers were later checked by exploitation, predation and introduced pathogens, whilst more resilient Eurasian grass species expanded (Crosby 1972). The Eurasian pig still is a pest in the woods of North America, where native pigs are lacking. The vegetation has not yet managed to adapt.

The rabbit in Australia: a classic introduction blunder

The introduction of the rabbit *Oryctolagus cuniculus* from Europe to Australia is a monumental blunder. The scale of its impact is considered to be unique in the history of exotic animal introductions.

The first attempt to introduce the rabbit in Australia was made by the British navy in 1788. The goal: to create a population of hunting game for meat and fur. The introduction failed, and further attempts followed, but all were unsuccessful until introduction on the south coast in 1859. The initial enthusiasm after a decade turned into concern when the Australians realised how much damage the rabbit was causing. In 1864, the rabbit was still legally protected for four months out of the year, but in 1875 the South Australian Rabbit Destruction Act was passed. Yet, in 1881 the first farms went bankrupt. By 1910, the rabbit had conquered most of Australia.

A summary of the damage to agriculture caused by rabbits:

- Heavy competition for grass with sheep and other livestock, resulting in fewer sheep with a lower production of wool, meat and lambs.
- Major damage to forestry, tree plantations, shrubs and grassland (perennials being replaced with less stable, annual plants).

The European rabbit, a major pest in Australia.

- Soil erosion, leading to silting of dams.
- Damage to agricultural and horticultural crops.
- Damage to forests and forest plantations.
- Impediment of the regeneration of forests and shrubs that prevent soil erosion in the arid interior of Australia.

Expressed in monetary terms, the costs amounted to some $600 million annually in agricultural damage by 1997, plus $20 million in pest control costs and another $5 million in research costs. The recent introduction of the RHD virus has now reduced these costs to about $200 million annually (see below).

The damage to biodiversity cannot be expressed in money:

- The threat to the survival of 17 indigenous species of plants, including 3 species of trees.
- The threat to the survival of various indigenous animal species like the greater bilby *Macrotis lagostis* (a burrowing marsupial), the Eyrean grass wren *Amytomis goyderi* and the plains-wanderer *Pedionomus torquatus* (a nocturnally active terrestrial bird).
- On Phillip Island: the extinction of the endemic parrot *Nestor productis.*
- On various islands: damage to the nesting places of breeding seabirds, including the endemic Gould's petrel.
- The increase in population size of predators such as the domestic cat (also an introduced species), causing additional danger to breeding seabirds.

From 1875 onwards attempts were made to exterminate the rabbit, with little success. Experiments with introductions of the myxoma virus, native to South America (where it targets the rabbit species *Sylvilagus brasiliensis*), started in 1936, but the first successful epizootic was generated only in 1950. A key factor were the mosquitoes transmitting the virus. Mortality among rabbits was 99% initially, but gradually dropped to 30-50%, as a result of co-evolution of the virus and the rabbit. Hence he number of rabbits as well as the amount of damage rose again.

In 1997-98, a new virus was introduced from China, the rabbit haemorrhagic disease virus (RHD). That proved successful. The virus has also decimated rabbit populations in Europe. But co-evolution is expected to occur in these cases as well.

Sources: Williamson (1997) and Bomford & Hart (in: Pimentel 2002)

Other Trojan horses among introduced productive species

There have been numerous well-intentioned introductions that turned out to be a disaster. From the long list of such introductions we give three examples.

The **gypsy moth** *Lymantria dispar* was brought from Japan to Europe in the 18th century. In 1869, it was imported into Massachusetts by a Frenchman who was seeking to crossbreed it with the silkworm. The goal was to produce a silkworm that not only produced high-quality silk but was also resistant to fungal and viral diseases. However, it turned out quite differently: the species developed into a destructive pest that caused massive damage to forests and ornamental trees, particularly oak. In 1981, some 3 million hectares were destroyed. Just like the oak procession moth, the species can cause skin irritation in humans. The caterpillars hitch on trucks to other states. Chemical pest control – first with arsenic, later with other substances – had little effect. The introduction of a pathogenic fungus, a virus and a bacterium also proved insufficiently effective. No effective natural enemy has yet been found (Veldhoen 1997).

Multiflora rose *Rosa multiflora* was imported from Japan to the US in 1886 for rose cultivation. With government support, the species was planted at a massive scale in the 1930s along the Eastern seaboard in order to combat erosion. Private companies marketed the species as a natural cattle fence and roadside shrub. In the 1960s biologists recommended it as feed and protective cover for wild animals. It was one of the few introductions considered beneficial by Elton in his classic 1958 book. However, it turned out that the species displaces indigenous vegetation, forms impenetrable monocultures and grows rampant, covering grassy fields and paths. From the 1960s onwards, it has been considered a pest; planting was prohibited and the species was combated with tractors, bulldozers, herbicides, goats, mites and imported wasps.*

The **golden apple snail** *Pomacea canaliculata* was brought from Florida and South America to Taiwan in the 1980s to set up snail farms. However, consumers showed little interest, and even less so when it was discovered that the species could transmit a species of lungworm. The snail farms were closed down and the snails were released, and spread into the rice fields, where they now cause some $1 billion in damage every year. In spite of biological control campaigns, using fish and ducks, the snail has already spread to Japan, China, Southeast Asia, Australia and Hawaii.**

* http://www.invasivespecies.gov/profiles/multiflorarose.shtml

** US Department of State, http://www.state.gov/g/oes/ocns/inv/cs/2312.html

The gypsy moth, a Trojan horse from Europe in North American forests.

Introduced hedges and ornamental plants becoming pests

Plants introduced as ornamentals or hedges are an increasingly frequent source of invasions. A few examples:

Rhododendron *Rhododendron ponticum,* was imported in the 19th century from Spain in the UK as an ornamental plant, and to give cover for pheasants. It has now become the nr 1 weed in woods and national parks on the British Isles, including Ireland. It displaces other plants in natural areas, including the endemic Lundy cabbage *Coincya wrightii* on Lundy Island. Control campaigns include herbicides and bashing by volunteers (Williamson 2002).

In South Africa many different trees and shrubs, including various **Acacia** species, were imported to serve as hedges, ornamentals etc. Many of these species now have to be actively combated (see Box 40).

The **common gorse** *Ulex europaeus* was imported into New Zealand in the early 19th century as a hedge plant. In the 20th century, it became the leading problem weed for farmers. It not only hinders livestock during grazing, but also displaces indigenous species in nature reserves (Williams & Timmins 2002).

The **common broom** *Cytisus scoparius* was brought to New Zealand to serve as an ornamental plant but has now become a pest in production forest and nature reserves, destroying open landscapes and threatening endangered plant species.

In the Netherlands, the **black cherry** *Prunus serotina* was introduced as early as the 17th century but did not develop into a pest until the 1930s, when it was planted to afforest heath fields. Control through bashing by volunteers and a parasitic fungus are only moderately effective.

The South American shrub **Lantana camara** was introduced in all tropical and subtropical regions, oftentimes to serve as hedge. The species is now a pest in forest plantations, orchards, agricultural land, grassland and natural areas. In Australia it is also penetrating indigenous forests. The species has a broad arsenal at its disposal. The leaves and fruits are toxic and can be lethal to livestock, birds and even children. The roots secrete a substance that prevents the seeds of other plants from germinating. The species increases the chance of fires, and benefits from such fires. In India it has even caused whole villages to be abandoned. Pest control is often difficult, costly, or labour-intensive, all the more because of the large variability of the species.

Another shrub from South and Central America, **Siam weed** *Chromolaena odorata (= Pelargonium odoratum)* was also introduced in tropical Africa and Asia as well as Micronesia. It was valued as an ornamental plant, but also as a competitor of *alang alang (Imperata cylindrica)*, an aggressive weed grass that often strikes after slash and burn in systems of shifting cultivation.

However, Siam weed also prevents forest from regenerating. It has colonized a broad range of habitats, ranging from farmland and plantations (rubber, oil palm, coconut palm, banana, forest and coffee) to riverbanks, shrub vegetations and (open) natural forests, often after disturbance. The species swamps, is flammable and thereby increases the chance of fires. Like *Lantana*, it is toxic to humans and other plants, and difficult to combat. Herbicides like 2,4-D and picloram are being used. Biological means of pest control has proven successful at a local level only.

The import of flowers and plants has become a major source of bio-invasions. In the Czech Republic, 53% of invasive exotic plants were imported as ornamentals. In Australia this percentage is as high as 65%. In Switzerland, 75% of all plants on the national black list were imported as ornamentals

In addition, these imports often do not come alone. In the US, 68% of insect interceptions at airports were found associated with the import of cut flowers. In Switzerland, 13% of imported cut flowers and nearly 5% of imported plants were contaminated with pest insects in the early 1990s.

Obviously, plants and flowers are high-risk trade flows.

Sources:
Ornamentals generally: Perrings et al. (2005)
Ornamentals Switzerland: Wittenberg (2005)
Lantana: http://www.bangor.ac.uk/~afs101/iwpt/web-sp6.htm
Chromolaena: http://members.lycos.co.uk/Woody PlantEcology/docs/ web-sp4.htm

Common gorse.

Lantana camara.

Co-evolution of pathogens and their human hosts after invasion

After an invasion, virulent pathogens and parasites immediately induce natural selection among their hosts, since in most host species there is intra-specific genetic variation in susceptibility to the pathogen. Resilient individuals have better chances to survive and reproduce. Simultaneously, the pathogens or parasites themselves undergo selection for less virulent strains. Most pathogens will evolve more rapidly than their host, since they can produce more generations in the same period of time.

A classic example of such a co-evolution is what happened in Australia after the myxoma virus was introduced to control rabbit populations. Mortality already started decreasing after a year (see Box 24).

Similar processes have occurred in human diseases that started as mass killer but ended as much less lethal childhood diseases. A disease that dies out because its surviving hosts have become immune but returns every five years or so, automatically becomes a childhood disease (McNeill 1976). Examples are **diphtheria**, **measles, mumps** and **smallpox**. In the case of humans, the process will take at least a century. The smallpox virus took several centuries before a less virulent type emerged, which was named alastrim.* Other diseases that became less mortal over time were influenza and the plague.

However, some human pathogens have not yet shown signs of becoming milder. It remains to be seen whether mitigation will occur and how long this will take.

The **plague bacterium** occurs in the Middle East and Europe at least since 1000 BC. During the Black Death (1346-1353) the case-fatality rate was about 80% for bubonic plague and 100% for pneumonic plague. These figures were still the same around AD 1900. This is explained by the fact that the flea infects its host only when it vomits blood into the wound. That happens only if it is infected with virulent plague Bacteria (Benedictow 2006).

In the early 20th century, the **cholera bacterium** developed a new strain, which became known as the El Tor strain. Patients who fell victim developed much *more* serious symptoms than those infected with the regular *Vibrio cholerae* bacterium. As yet, there is no sign that El Tor is mitigating towards a less virulent strain. In fact, there is no strong selection pressure in that direction: even if a patient dies, he or she has already secreted a significant amount of bacteria into the surface water which in most endemic areas is used by large numbers of people. This suffices for the bacterium to spread (Goudsmit 2001).

E. coli **0157** is a new strain of the common *coli* bacterium that has incorporated the cholera toxin. It probably came to Europe from the US in the early 1980s (Bol 2000). This zoonosis has already claimed thousands of deaths in the US. The carrier and transmitter are often cattle. Even vegetarians can fall victim, by drinking infected water or cider made from apples fallen to the ground in orchards where cattle graze. The most dangerous affliction by *E. coli 0157* is the Hemolytic Uremic Syndrome, which has a high case-fatality rate.

Infection by the **HIV virus** still has close to a 100% case-fatality rate if left untreated. It may take centuries before the virus grows less virulent. The AIDS epidemic has led demographers to adjust their prognoses for the global population in the third quarter of this century down by no less than half a billion.

Drug use will inevitably interfere with the selection process. First, it will often prevent genetic selection among hosts for better disease resistance, a consequence that is universally accepted in the case of humans. Second, it may limit selection in pathogens for less virulent strains, since hosts will live longer and can spread the virus longer in those cases where they continue to carry it during treatment (as in HIV). Third, drug use will induce selection among pathogens for drug resistance.

* Becoming milder didn't save the smallpox virus from global eradication in the late 20th century (see Box 46).

 # Introduced populations escaping from their enemies: the enemy release hypothesis

Species that have invaded a new area sometimes change in body-size, competitive force and density over time. They appear to have been unchained, and this is often true. Most invaders have been freed of at least some of their natural enemies: pathogens, parasites, herbivores or predators.

In their indigenous area, most plants and animals are often accompanied by numerous parasites. Tilden (cited in Elton 1958) gives the example of the coyote brush *Baccharis pilularis* in the western US. He found 257 species of arthropods on this shrub, including 221 insects. Of those, 53 were herbivores, 23 were predators and 65 were parasites (55 primary, 9 secondary and 1 tertiary parasite). Elton surmised that the success in Britain of an invasive species like rhododendron was possible because it has much fewer parasites there than it does in its native area.

Many different findings support this enemy release hypothesis. Recently, a large comparative study examined parasites of invasive animal species in their old and new area (Torchin *et al.* 2003). The study focused on 26 species of molluscs, crustaceans, fish, amphibians, reptiles, birds and mammals. The results:

- In their native area, the species have an average of 16 species of parasites; in the new area they have only 7.
- Of those 7 parasites, only 3 accompanied the species from the old area; the remainder are indigenous species.
- In introduced populations the percentage of infested individuals is relatively low.

The explanation might lie in the following "funnel":

- The population from which the species stems does not carry all parasites.
- The "random sample" from this population that is transported to the new area is small and therefore will be missing some parasites.*
- Some of the remaining parasites will not survive transport.
- Of the finally remaining parasites, some will die after settlement in the new area due to biotic or abiotic conditions in that area (e.g. a second required vector is missing in the new area).

Candidate parasites may also be present in the new area, but not all of those will be adapted to the new host.

As a rough rule of thumb: an invasive species has only half the number of parasites in its new area. Of those, half are old and half are new. For instance: in North America, the starling has only nine of its 44 European parasites.

Although the reduction in the number of parasites has been proven empirically, Colautti *et al.* (2004) warn that the now-popular enemy release hypothesis is used too easily to explain why invasive species are successful and to support the introduction of parasites. Their arguments:

- A smaller number of parasite species is not evidence of reduced parasitic pressure.
- The number of parasites is by no means always smaller than the number of parasites of *indigenous relatives* of the invasive species.
- The success of a species can be the result of human disturbance or a favourable climate.
- In the case of a deliberately introduced species, the initial population may have been *selected* for insignifi-

Red squirrel.

The grey squirrel is a reservoir of a parapox virus that kills red squirrels.

cant parasitism or may have been *treated* for parasites.
- A parasite can have a reduced effect on its host in the new area. For instance, it can sometimes come under attack from an indigenous predator. It can also prove more effective against an indigenous competitor of the invasive host than against the host itself. In that case, an invasive species can even benefit from the introduction of its natural enemy.

The authors advocate a case-by-case investigation of the factors involved.

However, the effect of parasites can also be *larger* than is assumed in the enemy release hypothesis (Prenter *et al.* 2004), as parasites can have other effects as well:
- They can increase their host's vulnerability to predators, for example by inducing behavioural change. That effect can be different for invasive and indigenous species.
- They can affect the competitive relation between an invasive species and an indigenous species. For instance, a parapox virus in Britain had little grip on the European red squirrel *Sciurus vulgaris* until the introduction of the Eastern grey squirrel *S. carolinensis* from North America. This latter species started to function as a reservoir for the virus, enabling it to become virulent for the indigenous squirrel as well. Likewise, the pheasant *Phasianus colchicus* introduced in Britain is functioning as a reservoir for a nematode that is a pathogen of the indigenous partridge *Perdix perdix*.

These authors, too, advocate restraint in the introduction of parasites and call for further research into the effects of parasites on species and biocoenoses.

In theory, an exotic species *A* from bioregion *P* could be more successful in a new bioregion *Q* than a related indigenous species *B*, which is a potential competitor. Conversely, species *B* from region *Q* could be more successful in bioregion *P*. If this is the case, it could be used by conservationists to set up a species exchange in order to save *both* species. But the situation will rarely be as simple and predictable as this. "Biota-engineering" is a tricky game to play.

* In this exceptional respect, a small initial population has a competitive advantage over a numerous initial population.

Introduced plants and animals swamping

Invasive species in some cases dominate whole areas, displacing indigenous or even other invasive species. Surprisingly, some of such swamping species take only a modest place in their native area.

As early as 1836, Darwin found that a crop imported from Europe to Argentina, **cardoon *Cynara cardunculus*** (of which artichoke is a variety), had formed impenetrable monocultures in Argentina and Chili, hundreds of square miles in size:

> "I doubt whether any case is on record of an invasion on so grand a scale of one plant over the aborigines."

Drooping brome *Bromus tectorum*, a grass species not quite common in Europe, has caused dramatic changes to the vegetation and fauna in some North American ecosystems. This annual, known as **cheat-grass** in the US, was first found in 1989 in British Columbia and has since spread widely over the shrub-steppes of Idaho and Utah. The grass makes these steppes more susceptible to fires. Before the invasion, a fire occurred every 60 to 110 years; after invasion, the frequency had increased dramatically to once every 3 to 5 years. The shrubs and accompanying rich flora were displaced by monocultures of cheat grass, which now occupy 5 million hectares, a surface area larger than the Netherlands. Songbirds, deer, antelopes, rabbits and rats dependent on the shrubs have also been reduced or disappeared entirely, followed by birds of prey (Bright 1998, Pimentel *et al.* 2005).

Purple loosestrife *Lythrum salicaria*, dominates in Europe on very few locations. It was introduced in the US as an ornamental. Today, it occurs in 48 states and swamps in many wetlands. Locally it reduced the biomass of 44 native plant species and endangers wildlife, including the bog turtle *Clemmys muhlenbergii* and several duck species depending on these plants (Gaudet & Keddy cited in Pimentel *et al.* 2005).

Yellow star thistle *Centaurea solstitialis*, native to Eurasia, has completely overrun 8 million hectares of grassland in California, resulting in the total loss of this once productive grassland (F.T. Campbell cited in Pimentel *et al.* 2005).

Common cordgrass *Spartina anglica* is a tidal zone plant. It developed around 1870 in Britain from a spontaneous crossbreeding between the indigenous *S. maritima* and the American *S. alterniflora*, which probably hitched a ride on a ship. The new species quickly spread across mudflats. Dutch researchers saw opportunities for coastal protection and land reclamation. After they published an article on this in 1929, orders for the grass flooded in to Britain from all over the world. The species was used to fix silt deposits, to reclaim land and to relieve waterways, but also to produce cattle feed and (in China) to produce green manure. It now occurs in Europe as well as all along North America's East Coast, East Asia, Australia and New Zealand.

Gradually, however, disadvantages became apparent too:

- Extensive mudflats with gentle inclines were replaced by fields of *Spartina* grass with steep inclines and poor drainage, intersected by deep canals.
- Some waterways became clogged due to fixation of silt deposits.
- The species not only thrived on mud flats, but also swamps other vegetation. In the Netherlands, it has almost completely displaced the indigenous *S. maritima.*
- Feeding areas for wading birds were lost.

In New Zealand, the fixing of silt deposits was so severe

Cheat-grass.

Purple loosestrife.

that drainage was hindered, leading to flooding. Indirectly, the species has contributed to land reclamation and thereby the loss of many salt marshes around the world. *S. anglica* does not yet have many natural enemies, partly because it is a new species all over its range.*

The **Japanese knotweed *Fallopia japonica*** was imported into Europe in the 19th century as a gardening plant. In Britain it is now considered the most important weed. The plant forms dense monocultures in clearings. In the Netherlands it is occurring in more and more areas along thickets, roadsides and railroad embankments. It has proven difficult to keep this plant under control. Even spraying with glyphosate is not very effective (Shaw & Seiger 2002).

The **water hyacinth *Eichhornia crassipes*** spread across the world from South America in the late 19th century after it was listed in a seed catalogue. Nowadays it is found in 50 countries on 5 continents. It can grow very fast, doubling its populations in as little as 12 days. It forms monocultures in all tropical areas, including Florida. It clogs up waterways, hinders shipping and creates a habitat for mosquitoes that transmit malaria and encephalitis. In addition, its shading and crowding of native aquatic plants dramatically reduces biodiversity in aquatic habitats.

Salvinia molesta, a floating aquatic fern native to Brazil, escaped from a botanical garden in Sri Lanka in 1939. The plant forms mats up to 1 meter thick. It has become a pest in large parts of Africa, India, Southeast Asia, New Guinea and Australia. In some areas, the damage even exceeded the damage caused by water hyacinth. For instance, the Sepik River in Papua New Guinea was clogged by *Salvinia*, which threatened the livelihood of 80,000 people dependent on the river. Fortunately, effective biological control methods have been found.

The **American or Atlantic jack-knife clam *Ensis directus*** grows in enormous densities on the seafloor in the tidal zone and the shallow sublittoral zone of the North Sea and is also inundating the beaches of the Netherlands. The species has marginalized indigenous species of razor and jack-knife clams (Reumer 2005).

Finally, another example from the Netherlands: the **Japanese (or Pacific) oyster *Crassostrea gigas*** was released into the Oosterschelde estuary in 1959, assuming that it could not spontaneously reproduce there. The species now covers more and more of the areas suitable for oysters and mussels and is also expanding into the Wadden Sea (see also Box 77).

Possible explanations for swamping are:
- Enemy release (see Box 28).
- The new area is (even) better suited for the species.
- The species is genetically more competitive than its indigenous competitors.
- The species produces chemicals harming indigenous competitors or natural enemies.

* www.nwcb.wa.gov/weed_info/ commoncordgrass.html

American jack-knife clam on a Dutch beach.

Japanese oyster swamping.

Ctenophores in the Black Sea: bio-invasion vs. bio-invasion

A single bio-invasion can radically disrupt a species community by becoming dominant and initiating ecological chain reactions. But sometimes, nature takes corrective measures in the form of a follow-up invasion of a natural enemy. A spectacular example is the introduction of the ctenophore **Mnemiopsis leidyi** in the Black Sea around 1980.

This species is indigenous to the Western Atlantic and probably arrived in Europe in ballast water. The first observations in the Black Sea were made as early as 1982. In the adjacent Mediterranean Sea it was signalled in 1992. The ctenophore fed not only on the zooplankton that formed the main food source for pelagic fishes like the anchovy, but also fed on the eggs and larvae of various fishes, including – again – the anchovy.

This ecological as well as economic disaster gave rise to a large amount of research, with the intention of finding a "remedy". The conclusion was that another ctenophore, **Beroë ovata**, could serve as an effective means of biological control. Remarkably, this species arrived spontaneously in the Black Sea in 1997 – probably also from America, and probably also in ballast water – and spread at a rapid rate. It feeds almost exclusively on *Mnemiopsis*. By 2001, it had almost completely eradicated the plague species. Subsequently anchovy catches went up again. So in fact, the global ecosystem has managed to solve this problem, albeit aided by humans.

It should be emphasised, however, that the *Mnemiopsis* invasion in the Black Sea took place in a period when the ecosystem was already disturbed. The anchovy population had been over-fished; the composition of the phyto- and zooplankton had changed in response to changes in atmospheric and sea currents in the Northern Hemisphere; and the sea was heavily polluted with nutrients, mainly from the Danube River. Subsequent reduction of the pollution from this river probably contributed to the reduction of *Mnemiopsis* densities.

The story continues, however. In 2001, *Mnemiopsis* appeared in the Caspian Sea, where it had already been observed in 1995. Probably it was introduced – again - with ballast water, this time through the Volga-Don canal. The effects were similar to those in the Black Sea. The invasion does not appear to have reached its peak yet. Some experts have suggested the deliberate introduction of *Beroë* as a biological means of control – provided all necessary preventive measures have been taken. This introduction will probably not take place until the reactions of the system can be sufficiently predicted – unless it again occurs spontaneously.

What makes experts careful is that the Caspian Sea and its biota are completely different from the Black Sea. Its isolated geographical location has produced a largely endemic fauna, which might be less resilient to species introductions.

In 2006, *Mnemiopsis* also exploded in the Baltic Sea. And it was found in high numbers in the Dutch Wadden Sea as well as in Lake Grevelingen.

Sources: Kideys et al. *(2001), Kideys (2002), Bilio & Niermann (2004), Tulp (2006) and* Volkskrant *(2006a, b).*

The first invader: *Mnemiopsis leydei*...

... and the follow-up invader that tamed it: *Beroë ovata*.

Two success stories of counter-introductions

Harmful invasive species are sometimes effectively controlled by the introduction of natural enemies. Elton (1958) used the term *counterpest* for this. The classic success story of such a counter-invasion stems from 19th century California.

Around 1868, the fluted or cottony scale insect *Icerya purchasi*, native to Australia, managed to settle in **citrus cultivation in California**. In search of a counter-pest, the **Vedalia ladybug *Novius cardinalis*** (also named Vedalia beetle and *Rodolia cardinalis*) was introduced, also from Australia. This resulted in a counter-invasion that proved effective within just a few years.

Later, the same method was successfully applied in Europe, South Africa, Japan, Hawaii, New Zealand and South America.

Those success stories led to euphoria over the ability to control bio-invasions using introduced natural enemies, and many other successes did follow.

A spectacular success was achieved in the 1980s in **cassava fields in Africa**, which were increasingly plagued by the mealybug *Phenacoccus manihoti*. Entomologists carefully searched for a monophagous enemy in the native range of cassava, Central and South America. They finally found one in Paraguay: the **parasitoid wasp *Epidinocarsus lopezi***. Large-scale breeding and introduction of this insect proved successful in large areas of Africa, often within a year following introduction

However, there were also many failures and blunders, as shown in the next Box.

Source: Elton (1958) and Zimmer (2000)

Vedalia ladybug, a succesful introduced counter-pest.

Failures and blunders in counter-introductions

The road to hell is paved with good intentions. This certainly applies to many counter-introductions in the past.

A classic example of a disastrous introduction is the **small Indian mongoose** *Herpestes javanicus (or H. auropunctatus)*. It was introduced to control rats on Jamaica, Puerto Rico and other **West Indian islands,** as well as on **Hawaii**, from 1872 onwards. Though the mongoose proved effective against the brown rat, it failed to control the black rat. In addition, it attacked ground-breeding birds and wiped out at least seven species of amphibians and reptiles. It also transmits rabies and leptospirosis.

The **house sparrow** *Passer domesticus* was introduced in the **US** in 1853 to combat the canker worm. By 1900, the species had become a pest causing damage to plants around buildings as well as to grain and fruit crops. In addition, it causes harm to indigenous birds, varying from the Baltimore oriole *Icterus galbula* and the yellow-billed cuckoo *Coccyzus americanus* to the bluebird *Sialia sialis* and cliff swallow *Petrochelidon pyrrhonota*. It also transmits 29 human and livestock diseases (Pimentel *et al.* 2005).

Also in the **US,** over 30 different **parasites** were introduced in the past century to combat the invasive gypsy moth. None of those parasites proved very effective, some even turned on a beautiful saturniid species, which is now threatened with extinction (Zimmer 2000).

The forests of **Hawaii** are full of **parasites** introduced from other regions in order to control insect pests. One example is a parasitoid fly introduced for controlling a species of shield bug. It turned out that this fly also attacked *Coleotichus blackburniae*, a large and colourful endemic bug, which has now virtually disappeared. Another example is parasitoid wasps introduced to control moths that cause damage to agricultural crops. The wasps also attacked a variety of indigenous species, causing Hawaiian forest birds to lose an important source of food (Zimmer 2000).

Also on **Hawaii**, the **starling** *Sturnus vulgaris* was introduced to control cutworms and armyworms that threatened sugar cane crops. Instead, the species started to actively distribute an invasive weed species (Pimentel *et al.* 2005).

Particularly foolish proved the introduction of the **rosy wolf snail** *Euglandina rosea,* a native to Florida and Central America, on **Pacific islands** in the 1950s. It was introduced to control the previously introduced giant East-African land snail *Achatina fulica* that caused damage to crops and gardens. Though the predator did attack the giant snail, it hardly affected its population size. Instead, it became a disaster to biodiversity. It wiped out 30 endemic species and subspecies of tree snails within just 30 years, particularly on the Society Islands. Most belonged to the Partulidae family, which in terms of adaptive radiation can be compared to Darwin's finches of the Galapagos Islands and the Drepanididae of Hawaii. The wolf snail is expected to wipe out all remaining endemic tree snails in the Pacific as well (Williamson 1997).

The **cane toad** *Bufo marinus,* native to South America, was imported from Hawaii to **Australia** in 1935 to combat beetles that were damaging sugar cane. The toad proved not very effective against the beetles, but spread widely across large swaths of Northern Australia, where it now occurs in higher densities than in its indigenous area. The species has proven harmful: it is highly toxic in all stages of life and kills a large numbers of other animals, including dogs, frogs and even crocodiles. It also predates on honeybees and indigenous species and transmits diseases to frogs and fishes (see also Box 33). Control attempts have failed so far. Plans are being made for large-scale biological control using genetically modified (!) toads (Groeneveld 2006).

Occasionally, natural enemies are introduced to combat *indigenous* pests. For instance, the **mosquito fish** *Gambusia affinis* from North America was introduced in **numerous countries** to combat mosquitoes (as well as for anglers). It is now probably the world's most widely distributed freshwater fish (Elton 1958). It is a pest, is harmful to native species and has crossbred with one endemic species: *G. heterochir* (Williamson 1997).

Most cases mentioned above refer to polyphagous introduced predators and parasites. However, even monophagous biological enemies can have adverse side effects. An amazing example was recently discovered in **Western North America** (Pearson & Callaway 2006).

Two **gall flies of the genus *Urophora*** were introduced
there in the early 1970s to control the exotic spotted and
diffuse knapweed *Centaurea maculosa* and *C. diffusa*,
which aggressively invade arid habitats. It appeared that,
although the flies managed to reduce seed production in
C. maculosa, they did not effectively control its numbers.
The weed continued to spread to new locations and
increased in abundance. The fly larvae, however, were
welcomed as an additional food source by deer mice of
the species *Peromyscus maniculatus*, which thereby
could double its densities. Unfortunately these deer mice
are a reservoir for several human pathogens, including
the Sin Nombre hanta virus, which causes hanta virus
pulmonary syndrome (HPS), a disease killing about a third
of its victims. This food chain can be summarized as follows:

> weed > herbivore fly > deer mouse > hanta virus >
> humans

This is how a monophagous fly introduced for weed control can promote a human disease.

The cane toad has become a major pest on Pacific islands and in Australia.

Escalating counter-invasions in Micronesia: going from bad to worse

The cure of a counter-invasion can be worse than the disease, and it can even provoke additional counter-invasions. A shocking example of this is provided by an escalating series of counter-invasions in Micronesia.

It started long ago with the introduction of **rats** on the islands. To control the rats, the **monitor lizard *Varanus indicus*** was introduced. That proved not a good idea. Rats are active at night; they are not suitable prey for the diurnal monitor lizards. So the monitors turned to poultry instead.

Sometime before 1945, the **cane toad *Bufo marinus***, native to South America, was introduced to give the monitors something else to eat and perhaps also to keep down insects in the coconut plantations. The toad secretes a powerful venom from its skin, and large numbers of monitors were poisoned. As the monitors declined, one of the coconut pests, a rhinoceros-beetle, underwent a population explosion, because the monitors had eaten its grubs. With the monitor out of their way, the toad population could further increase. Cats, dogs and pigs attacked the toad and were killed in turn. That was good news for the rat, which could explode.

The **giant East African land snail *Achatina fulica***, brought in by the Japanese during World War II as a food source, exploded as well, perhaps partly because of all the available carrion in the form of dog and cat carcasses. To control the snail, the next counter-pest was introduced during the 1970s and 1980s: a **predatory flatworm**. The flatworm is currently spreading throughout the islands and has become a major new threat to Oceania's extraordinarily diverse native snail fauna.

Summarising, there was one accidental introduction (rat) followed by four intentional introductions:
- a counter-pest (monitor lizard) was a partial failure;
- a food species (the giant African snail) became a pest;
- a "feed" species (cane toad) to distract the lizard became a pest itself and generated additional pests (beetle and rat);
- a "counter-counter-pest" (flatworm) became a pest, too.

All 5 introduced species were polyphagous and all became pests, the 4 intentional introductions no less than the initial accidental introduction.

Source: Bright (1998)

Giant African snail.

Nature reserves are no safe havens from bio-invasions

Nature reserves can offer protection against a wide range of threats, ranging from hunting and pollution to habitat destruction. But they are defenceless against climate change and it would be naive to think that reserves are safe havens from bio-invasions. A study of **23 nature reserves across the world** found that 18% of terrestrial vertebrates and 30% of vascular plants are exotic. These percentages correlate with the number of park visitors (Usher 1988), which in turn correlates with propagule pressure and perhaps with habitat disturbance.

Not surprisingly, the most dramatic invasions occur in reserves on islands, particularly oceanic islands:

- In **New Zealand**, over half the 2000 reserves require weed control efforts (Williams & Timmins 2002).
- In reserves on **Hawaii**, 50% to 70% of all vascular plants were found to be exotic (Vitousek *et al.* 1997).
- On the **Galapagos Islands**, the number of exotic animal species has risen to 785, including some that were introduced before the park was created. The impact is dramatic. Of the islands' 11 native species of bats and rats, 8 are now extinct, 6 of them because of exotic pressure. Wild pigs are eating the eggs of the famous giant tortoises, the endangered green sea turtle and several iguana species. On Pinzon Island, black rats kill virtually every hatchling of the local race of giant tortoise. On some islands, goats have eliminated all the seedlings of several native tree species, and feral house cats have eaten most of the lava lizards. An exotic ant, the little fire ant *Wasmannia auropunctata*, has suppressed most of the islands' native ant species (Bright 1998). Even the world-famous and strictly protected Galapagos finches are now under threat from accidentally introduced parasites (*Der Spiegel* 2004). Plant invaders, too, are advancing. The shrub *Lantana camara* threatens endemic vegetation and breeding birds. And on several islands, a major bramble invasion has begun to take place.

- Even on **Java**, a continental island, nature reserves are being invaded. *Acacia nilotica*, native to Africa and India, was introduced in Baluran National Park as a fire-resistant tree. Unexpectedly, the tree began to spread across the treeless grassland at breakneck pace. This changed the landscape and adversely affected the living conditions for endangered animals, including the wild banteng *Bos javanicus*. The acacia is hard to control and is resistant to grazing as well as fire and drought (www.bangor.ac.uk/~afs101/iwpt/web-sp1.htm).

However, nature reserves on **continents** are also increasingly being invaded. For example:

- Alien weeds are invading approximately **700,000 ha per year of the US wildlife habitat**.
- In the **Great Smoky Mountains National Park** in the US, 400 of 1500 vascular plants are exotic (Pimentel *et al.* 2005).
- Nature reserves in **the Netherlands** have been invaded by such exotic trees as black cherry *Prunus serotina*, red oak *Quercus rubra* and sycamore maple *Acer pseudoplatanus* (some of them, however, planted).

South-African student studying invasive Lantana shrubs.

5 Impacts on nature, biodiversity and environmental quality

Biodiversity generally

Bio-introductions can be considered a "short circuit" between two separate biological networks. They create new ecological interactions, which are often hard to predict.

By definition, bio-invasions have both a positive and a negative initial effect on biodiversity. The positive effect is that the recipient area gains a new species.[50] The negative effect is that the difference between that area and the native area of the invader declines.
What happens next is not quite as clear-cut. Will the invasive species displace any native species? If not, then the local increase in biodiversity will be permanent. But if yes, biodiversity will drop again locally. If this causes an *endemic* species to become extinct, there is an *irreparable* loss to global biodiversity.

Such extinction as a result of bio-invasions has, to our knowledge, never yet been found in oceans and on continents, though it has been frequently reported on islands, in lakes and in rivers. But even on continents, invasive species have been a decisive factor in the dramatic decline of many species, occasionally to the edge of extinction. Now that the frequency of bio-invasions is increasing globally (Welcomme 1992), bio-globalisation is becoming an escalating hazard for the diversity of species and biotas.

The share of bio-invasions in the decline of global biodiversity is currently under discussion in the scientific literature (see Box 41). According to one source, bio-invasions are now the second-most important cause of biodiversity loss, after habitat loss, in the US. The IUCN ranks bio-invasions as the third important cause after habitat loss and exploitation. One other study, focusing on the future, considers bio-invasions the fourth important threat after habitat loss, climate change and nitrogen (NO_x and ammonia) deposition but with a key role in islands, lakes, waterways and Mediterranean areas.

However, there are significant differences from one species group to another. The "winners" include groups such as grasses and ducks, among the "losers" are cactuses, parrots and monkeys (Sax & Gaines 2003).
To complicate things, there are also contrasts between geographical scales. The *decline* in biodiversity on a global scale (partially as a result of bio-invasions) is often coupled with a biodiversity *increase* on a regional scale (as a result of the same bio-invasions). Thus we are facing a paradox of scale, which can lead to confusion.

Biodiversity on continents

On continents, chances are slim that a species will suffer extinction due to bio-invasions.[51] To our knowledge, only a single invasive species has managed to wipe out other species on continents: *Homo sapiens*. And this has happened on a large scale. Hunting humans were, most likely, a primary cause of the mass extinction of large mammals at the end of the Pleistocene (see Box 35). In historic times humans have also wiped out numerous species of plants and animals – either directly, through hunting and gathering; or indirectly, through destruction of habitats.

Perhaps the most baffling example of this is the extermination of the passenger pigeon *Ectopistes migratorius*. In the early 19th century, this species still occurred in overwhelmingly large numbers: flocks darkened the sky and by 1810 a single flock was observed with an estimated 2.2 billion individuals! However, by 1900 the species had disappeared from the wild, and in 1914 the last individual died in captivity (Quammen 1998).

Some 400 (42%) of the 958 listed endangered species in the US (including Hawaii and other areas) are primarily threatened by invasive exotic species (Pimentel *et al.* 2005). The American chestnut *Castanea dentata*, which once constituted 25% of deciduous forests in the US, is now all but extinct due to a fungus introduced from Asia, the chestnut blight fungus *Cryphonectria parasitica*. The fungus was introduced from Japan in the late 19th century when it hitched a ride with a cargo of Japanese oak, which itself is resistant to the fungus (Anagnostakis 1997).

Further examples of near-extinctions on continents due to bio-introductions:

- In Australia, the rabbit is threatening one species of marsupial, 2 species of birds and 17 species of plants (see Box 24).
- The rinderpest virus, introduced from Asia in 1889 with a shipment of livestock, has caused

The American chestnut has been near eradicated by a fungus accidentally introduced from Japan.

mass mortality among ungulates and their predators; it has almost completely eradicated the Cape or African buffalo *Syncerus caffer.*

- South Africa has a wealth of flora, with 21,137 species, of which no less than 80% are endemic. Many of those are endangered. In the best-studied and most frequently invaded biome, the *fynbos* in the Cape Province, 80% of cases involve threats by exotic trees, particularly *Acacia*, *Hakea* and *Pinus* species (see Box 40).

Some invasive species are even able to alter ecosystems. Grazing mammals, for instance, can inhibit regeneration of forests or cause erosion. Plants can introduce extra nitrogen into the environment. Other plants may extract large amounts of water, causing drought and raising the frequency of forest fires (see Boxes 40 and 42). Recent examples are the large-scale forest fires in Portugal in the summer of 2005, particularly in and around plantations of *Eucalyptus globulus*, a species that absorbs and evaporates enormous amounts of water and easily catches fire.

Biodiversity on islands

Indigenous species of islands are often ecologically naive: they are adapted to only a small number of pathogens, predators and competitors and have lost some or most of their defence mechanisms. That makes them extremely vulnerable to invasive species.

On numerous islands, the endemic fauna and flora have been ravaged by the intentional introduction of:
- carnivores: cats, dogs, mongooses and snails;
- herbivores: sheep, goats, deer, rabbits and monkeys;
- omnivores: pigs.

Goats and sheep were even introduced on many small, distant and uninhabited islands as a food source for survivors of shipwrecks. To this must be added numerous unintentional introductions of rats, mice, ants, snakes, parasites and pathogens.

Of all known extinctions since AD 1600, 75% have taken place on islands (Bright 1998). To name but a few of the many islands where such tragedies have occurred: St. Helena in the Atlantic Ocean (see Box 36); Christmas Island in the Indian Ocean (Box 37) and Guam and Hawaii (Boxes 38 and 39) in the Pacific.

Darwin remarked in *On the Origin of Species* (1859):

"In many islands the native productions are nearly equalled or even outnumbered by the naturalised; and if the natives have not been actually exterminated, their numbers have been greatly reduced, and this is the first stage towards extinction."

The honeycreepers of Hawaii show spectacular adaptive radiation. Many species went extinct due to introduction of the avian malaria parasite and its mosquito vectors (From: Pratt *et al.* 2005).

The most destructive effect of invaders is predation. Simberloff (cited in Williamson 1997) analysed 71 unequivocal cases of island extinctions resulting from invasions, finding that 50 of these extinctions were caused by predation; 11 by destruction of habitats by the invader; and only 3 by competition.

For birds, fairly complete global figures are available (Williamson 1997). Since AD 1600, at least 93 species and 83 subspecies have become extinct; many other species are endangered. The most important causes are:

- Habitat loss was involved with 19% extinct bird species and 58% of those endangered.
- Hunting with 15% extinct and 26% endangered birds.
- Predation (mainly by introduced invaders) with 42% extinct and 40% endangered birds.

So predation was the most important factor in these cases, and may well continue to be a key factor in the future. On Pacific islands the three main threats to the relict avifauna are introduced species, habitat loss and over-hunting. Most threatening are introduced predators, mainly rats, cats, small Indian mongoose and the brown tree snake (Millet 2006).

Parasites and their vectors, too, can have a dramatic influence, as was shown by the introduction of mosquitoes on Hawaii that could transmit avian malaria, causing the extinction of many endemic bird species (see Box 39).

To mention two other examples of extinctions and possible extinctions on islands: goats introduced on San Clemente Island near California have wiped out eight endemic plant species and are endangering eight others (Pimentel *et al.* 2005). On Hawaii, where 35% of the 2690 plant species are exotic, no less than 800 (46%) of the 1744 indigenous species are endangered (Pimentel *et al.* 2005).

Thus on many islands a silent battle is going on between introduced and indigenous species, a battle that recently even figured in *The Economist*, a magazine that strongly advocates the blessings of economic competition.[53]

Biodiversity in lakes

Finally, we mention some examples of the effects bio-introductions may have on biodiversity in lakes:

- Following the introduction of the Nile perch in Lake Victoria, more than half of the original 350 species of fish – over 80% of which are cichlids – were driven to extinction or reduced to a minimal population (Goldschmidt 1994). In addition, the flesh of the fish is oilier than that of the local fish, so more trees were felled to dry the catch. The subsequent erosion and run-off contributed to increased nutrient levels, opening the lake up to invasions by algae and water hyacinth. These invasions in turn led to oxygen depletion in the lake, which resulted in the death of fish (ISSG undated).
- Other destructive effects of introduced species - including predation, competition, disruption of food webs, and introduction of diseases - are known from the Cuban freshwater systems, lake Titicaca in Peru and Bolivia, and lake Atitlán in Guatemala (Revenga & Kura 2003).
- In China, the introduction of tens of exotic species was found to cause major changes in the species composition of the fish fauna, with some species gone extinct. This included Lake Dian Chi (which had many endemic species) and Lake Dong Hu (Xie *et al.* 2001).
- In North America, the construction of the Welland Canal in 1829 enabled the sea lamprey *Petromyzon marinus* to circumvent the barrier of Niagara Falls and penetrate into Lake Erie and other lakes. It took a century before the effects of this became clear: three endemic fish species became extinct, causing major damage to fishery (Elton 1958, Simberloff 2000).

In tropical Asia, herbivorous and omnivorous fish (such as various species of carp) were involved in the vast majority of introductions. However, these species from temperate climates have never represented more than a minor share of fish catches in this region. The largest impact appears to be caused by introductions of carnivorous fish from temperate climates (trout and bass). In many cases the goal of the introduction – better yields for commercial fishing – was achieved. But in some cases the introduced species began to dominate the fish catches in a particular lake, reflecting that the composition of species had undergone a radical change.

Similar phenomena as observed in islands and lakes have occurred in other isolated bioregions, such as mountains invaded by species from other mountain areas.

Mechanisms

Which ecological mechanisms can lead to the displacement of a species? There are at least four:

1 Predation: the invasive species predates on a native species but, being polyphagous, remains

After its introduction in Lake Victoria, the Nile perch displaced some 200 endemic fish species.

numerous after the native species has become rare.

2   Parasitism: the invasive species is (or spreads) a pathogen or parasite against which the native species has no natural defences.

3   Competition: the invasive species competes with a native species. This is known as "amensalism" when the competition is unilateral. In some extreme cases, the invasive species can take over an entire area or dominate it in some other fashion ("swamping").

4   Indirect effects, e.g., by affecting the habitat of native species (see Box 42).

Last-resort remedies

On various islands, rescue operations to save endemic species are ongoing. In some cases these only delay the inevitable extinction, but in other cases they are successful, at least for the time being. On Mauritius, for instance, the endemic kestrel was saved from extinction at "five minutes to midnight" (see Box 49).

In some extreme cases, the last few individuals of a species were caught and held in captivity. The individuals or their offspring may then be released elsewhere or in their native habitat when conditions have become safer. In some cases predators are eradicated for that purpose. A recent example is a programme to save the endemic flightless Campbell Island teal *Anas nesiotis* of Campbell Island, some 700 km south of New Zealand (see Box 49).

Effects on environmental quality

Some introduced species have a direct effect on environmental quality, by inputting nitrogen and depleting aquifers. Many more invasive species cause indirect effects since they are controlled using chemicals.

During export of plants and plant products like wood, the battle on invasion of pest species often begins before the material even leaves the exporting country.[54] This is because the exporter seeks to avoid trade bans. Highly toxic broad-spectrum decontaminants are often used for this. These increasingly include methyl bromide, in spite of international agreements to phase out ozone-depleting substances.[55]

When a harmful species has nevertheless managed to penetrate another bioregion, attempts are often made to eradicate, contain, or control it, using chemical, mechanical or biological methods. Globally, a significant amount of the pesticides deployed in agriculture and forestry are used to combat invasive exotic species. This is because exotics have a high share in pest and weed species. In US pastures, 45% of the weed species are exotics, and in agriculture even 73%. In California, two-thirds of all damage to crops is caused by exotic insects and mites (Pimentel *et al.* 2005). Of the 40 top crop pests in South Africa, 42% are exotics (Lach *et al.* 2002). In New Zealand as much of 90% of harmful invertebrates are exotics (Cook *et al.* 2002). And in Brazil, pesticides used for control of exotic species are causing large-scale environmental pollution (see Box 48). In fact, since agriculture in most countries is based primarily on exotic species, and many plant pathogens and pests originate in the native range of the plant, we may safely assume that at least half of the global pesticide use is associated with introduced pests.

Some examples of chemical control outside agriculture:

- In the past: the use of DDT and dieldrin in Australia to combat ants;
- More recently: the use of insecticides in the US against *Aedes* mosquitoes transmitting the West Nile virus.
- Extermination of rats, possums, rabbits, deer and chamois in New Zealand using toxins, in part to protect endangered species of animals.
- Use of the herbicides 2,4-D and picloram in tropical countries to combat the shrub *Chromolaena odorata*.
- Use of chlorine and copper sulphate against the zebra mussel in the US and the black-striped mussel in Australia.
- Use of pesticides against the sea lamprey, a parasite of fish, in the Great Lakes of the US.
- Use of glyphosate (Roundup) against water hyacinth and Spartina grass.

In the case of a pathogen, if control of the pathogen or its vector has failed, the last resort is to try and make the plant or animal resistant to the pathogen. In animals vaccination can be an effective method. In plants we have to rely on breeding. Using the classical breeding methods, this is normally a lengthy process. Genetic engineering may offer a fast track, but may create new biological or commercial risks. For instance, in France a discussion is ongoing about a race of grapes made resistant, using genetic engineering, to the nematode that causes fanleaf disease. Opponents fear this will cause damage to the reputation of French wine.[56]

Very few attempts have been made to express the effects on biodiversity and environmental quality in monetary terms. In fact it is hardly possible to do so.

Prehistoric extinctions from invasions of *Homo sapiens*

In terms of biodiversity, *Homo sapiens* is perhaps the most destructive invasive species in the history of life. The two largest mass extinctions since the Ice Ages are probably largely due to humans: birds on islands in the Pacific Ocean and large mammals on various continents.

Mass extinctions of **large mammals on all continents** occurred towards the end of the Pleistocene. These mammals included the mammoth, mastodon, woolly rhinoceros, Irish elk, sabre-tooth tiger, several giant sloths and various large marsupials. The figures differ from one continent to the next. The proportion of genera that suffered extinction was:

- in North America: 75%
- in South America: 76%
- in Europe: 45%
- in Australia: 43%
- in Africa: 13.5%.*

Leaving Africa and Asia aside, there was a spectacular correlation between body size and the percentage of extinct genera. Percentages for different body weight classes were:

- <5 kg   1.3%
- 5-100 kg   41%
- 100-1000 kg 76%

Scientists are still debating the exact causes of these extinctions. The rapid climate change at the time seems to offer a good explanation, but such climate changes had occurred previously without causing mass extinction.** Hunting is a more likely explanation, but in the eastern US traces of hunting have been found in just three species: bison, mammoth and mastodon. Perhaps domino effects played a part, as predators switched to other prey species as bison, mammoth and mastodon became rare. In addition, these three species may have played a key role for other species by maintaining a varied habitat. But in both cases, mass extinctions can be largely attributed to a single invasive species: *Homo sapiens*.

As for **islands in the Pacific Ocean,** wherever Polynesians settled many **bird species** soon became extinct. For instance, there were 12 species of moa (giant flightless birds) on **New Zealand**. Between AD 900 and 1600, all these species were driven to extinction due to hunting, gathering of eggs, habitat destruction and fires. A further 15 species and subspecies of birds also suffered extinction, including rails, a swan, geese, ducks, a pelican, a crow and other birds of prey as well as a nightjar. Most of these species lacked the ability to fly and were active during day-time. Three species of nocturnally active kiwis survived the massacre.

On **other Pacific islands**, at least 1600 bird species became extinct within just a few centuries following the settlement of humans. That is a huge number compared to the estimated 9000 bird species that remain. On Hawaii alone, 50 species went extinct, including two ibises, four ducks, six birds of prey and owls, various rails, various crows and dozens of the most characteristic group of birds on Hawaii: the Hawaiian honeycreepers Drepanididae.

The Polynesians also brought dogs, pigs and the Polynesian rat with them to Hawaii, and these no doubt also caused significant damage. We were lucky that the Polynesians did not colonize the Galapagos Islands, where 95% of the original species still survive.

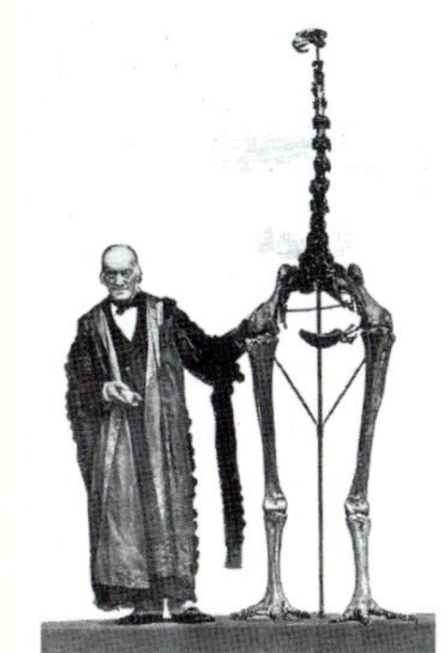

Richard Owen next to the skeleton of the tallest moa, *Dinornis giganteus*.

* Humans can only be considered exotic beyond Africa. The relatively low percentage of extinctions in Africa may be related to the lengthy co-evolution of humans and large mammals on that continent (Diamond 1999).

** However, a recent article (Shapiro *et al.* 2004) indicates that the genetic diversity among bison had already plummeted before humans colonized America, indicating that climate change might have played a more important part.

Source: Williamson (1997)

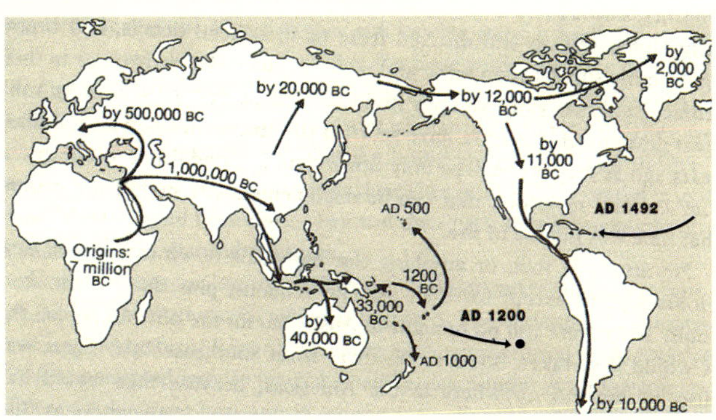

The distributional history of *Homo sapiens* (after Diamond 1999, modified).

St. Helena: how introduced goats destroyed unique island life

St. Helena in the South Atlantic is well known as the island where Napoleon was exiled, died and was buried. Few people know that previously it had become the grave of many endemic plant and animal species, which still lack a gravestone.

St. Helena was covered with lush forest when it was discovered in AD 1502 by the Portugese. They soon introduced goats and pigs on the island. The goats reproduced rapidly, feeding on the forests' young growth, thereby hindering forest regeneration. The Portugese, the British East India Company and the Dutch East India Company (VOC, see Boxes 72 and 73) used the island for victualling their fleets. The VOC abandoned St. Helena in 1652 as a stopover point in favour of the Cape of Good Hope in South Africa.

In 1659 the British East India Company took over the island. They began to exploit the adult trees. Redwood and ebony trees were used for tanning and for burning limestone for the construction of fortifications. Soon the forest was cleared but for steep slopes and gorges. Rainwater had already washed away the fertile soil.

In 1709, the governor alerted the directors of the Company, asking permission to kill the goats in order to save the ebony forests and combat the increasing drought. The reply was very clear:

"The goats are not to be destroyed, being more valuable than ebony."

Not until 1731 was it decided to rid the island of goats; but by then, the damage had been done. The forest was gone, and wood had to be imported at high cost.

In the meantime, numerous plants and hunting game species had been introduced from Europe, America, Australia and South Africa. As these species spread, they displaced indigenous flora and fauna.

By the 1830s, habitat destruction and introductions combined had resulted in the extinction of numerous endemic species – and even genera – of land snails, seed plants and ferns.

Sources: Darwin (1836) and Wallace (1880)
Website.lineone.net/-sthelena/history.htm

Christmas Island: endemic land crabs massacred by an invasive ant

Christmas Island in the Northeastern Indian Ocean has the world's largest population of land crabs, belonging to three species. They play a key role in the forest ecology by eating seedlings and leaves, digging burrows, turning over the soil and fertilising it with their droppings.

Between 1915 and 1934, the yellow crazy ant *Anoplolepis gracilipes* was accidentally introduced on the island from Africa or Asia. In 1995, it formed super-colonies and became a disruptive pest. The ant killed 20 millions crabs. With the crabs gone, seedlings and saplings could form a carpet; and the giant East-African land snail could invade the forest. The invader protected honeydew-producing scale insects from their enemies, allowing these insects to explode and kill forest trees. This created light gaps in the canopy encouraging seedling growth and weed invasion into the forest. The ant also threatened endemic arthropods, reptiles, birds and mammals; and it became a pest in houses and agriculture.

In 2002, a campaign was launched in which a helicopter spread baits contaminated with the insecticide fipronil. Though the campaign quickly decimated the ant

Yellow crazy ant.

population, it did not entirely eradicate it; the ant could even partially recover and spread. Biological control methods are now being developed.

Sources:
www.latrobe.edu.au/news/2004/mediarelease
www.issg.org/database/species/ecology.asp?si=110&fr=1&sts
www.banksiafdn.com/page_assets/Case2003_crazyant.pdf

The brown tree snake on Guam: a single invasive species destroying endemic biodiversity

Originally, the island of Guam in the Pacific Ocean consisted mainly of forest. The 22 indigenous bird species were forest, marsh and sea birds. They were joined by 8 introduced species, particularly in man-made landscape and built-up areas. In the forest, there were 12 native species and one exotic species. Snakes did not occur.

A radical change of the scene was initiated around the year 1950, when a military transport ship arrived with a stowaway from the Admiralty Islands: the brown tree snake *Boiga irregularis*. Initially, this nocturnal snake was hardly noticed. In 1970, an indigenous reed warbler became extinct, possibly due to the snake. Since then, 9 of the remaining 11 indigenous forest bird species have suffered extinction. In addition, 2 out of 11 indigenous lizard species became extinct.

By contrast, the eight introduced bird species all managed to survive. Two indigenous herons also survived, either because the snake did not penetrate into their breeding biotope or because the herons managed to defend their nests.

Two of the remaining bird species, a rail and a kingfisher, have been captured, and survive in captivity. But their fate, too, seems sealed because no method to eradicate the snake has yet been found. Guam is now close to an ornithological desert. The forests are full of spider cobwebs. The snake still survives on a diet of lizards.

Alarming enough, the snake was also observed in harbours of other Pacific islands including Hawaii. It has even been spotted in airplanes. Thus far, the invaders have been intercepted, but one day a snake may slip through and start a new massacre.

Sources: Quammen (1996) and Williamson (1997)

Brown tree snake, killer of endemic birds and reptiles on Guam.

Hawaii: killing fields of endemic species

The main battlefield of invasive species versus indiginous birds since paleolithic man are the Pacific Islands. Number one is Hawaii. Hawaii is an archipelago of oceanic islands. It is located at a very long distance from Asia and America, creating the conditions for an exceptionally unique indigenous flora and fauna:

- 1126 Species of flowering plants, 91% of which are endemic. Particularly remarkable are the tree ferns that reach a height of 8 metres, violets reaching a height of 2 metres, and over 100 species of *Lobelia*. There are even *Lobelia* species shaped like shrubs, climbing plants and small trees.
- 111 Species of endemic breeding birds, including the spectacularly colourful and diverse Hawaiian honeycreepers (Drepanididae). The 50 species and subspecies in this family have a wide variety of beak shapes, many of them adapted to the species of *Lobelia* foraged.
- A bat, but no terrestrial mammals.
- Sea turtles, but no terrestrial reptiles.
- No amphibians.
- Six freshwater fishes, all of them endemic.
- Over 800 species of land snails, more than anywhere else in the world. Of those, 99% are endemic.

Over the centuries, the flora and fauna of Hawaii have been ravaged, first by Polynesians and from AD 1778 onwards also by Europeans. Initially, the main threat was from hunting; later, other mechanisms became equally important:

- Habitat loss due to agriculture, forestry and the construction of homes.
- The introduction of pets and livestock including pigs, goats, dogs and cats, as well as a giant snail. Pigs, rats and cats have wreaked havoc (and continue to do so) on birds and their eggs. Pigs also cause damage to the soil and shrubs and spread seeds of exotic plants. Goats feed on the vegetation, leaving the dry slopes barren.
- The introduction of the Indian mongoose and the barn owl to combat the previously introduced rats. These predators, too, wreaked havoc on birds or their eggs.
- The introduction of a predatory snail to combat the introduced giant snail (see also Box 32). This led to massacre among the endemic snail species.
- The accidental introduction, in 1827, of mosquitoes from Mexico. They provided the vector that the previously introduced avian-pox virus and avian malaria parasite had been waiting for. The mosquito transmitted these pathogens from introduced fowl to native birds. Before the close of the 19th century, 16 species and subspecies of the magnificent Drepanididae family had suffered extinction.*
- The extinction of pollinating birds and insects led to greater and greater reproductive problems for the indigenous flora.
- Even the remaining indigenous forests are now full of non-pollinating exotic insects; and an exotic ant species is feeding on pollen.
- Invasive trees, shrubs and climbing plants are displacing indigenous species.

The drama continues. Of the 111 indigenous species of birds, 51 are extinct and 40 are endangered. Of the native 1126 flowering plants, 93 have gone forever and 655 are under threat. With only 0.2% of the total surface area of the US, Hawaii has become the world's "extinction capital".

Source: Bright (1998)

Note: just before we finished the text of this book, BBC news reported that avian malaria had killed the presumably last surviving individual of the brown *po'o-uli* (*Malamprosops phaeosoma*), another member of the Drepanididae family. It had been captured and held in captivity for a breeding programme intended to save the species from extinction. The species was first discovered as late as 1973. (see: *news.bbc.co.uk/1/hi/sci/tech/4066077.stm*)

The *'o'u (Psittirostra psittacea)* was one of the magnificent honeycreepers of Hawaii.

Invasions in a botanical treasure house:
the Cape Province

The Cape Province in South Africa has a unique and exceptionally rich flora counting between 8000 and 12,000 species, most of which are endemic.* This flora is even considered one of the world's six great flora realms. The richest of all is the *fynbos* in the southwest, a real botanical treasure house. This flora is slowly but surely being invaded and displaced by exotic plants.

The process started in the late 17th century, when Huguenots from the Mediterranean region imported the **pine-tree *Pinus pinaster***, which grew well in the Mediterranean climate of the Cape Province.

It was followed in the late 18th century by the **prickly pear *Opuntia ficus-indica***, a cactus used for ornamental purposes, as well as for separating parcels of land and for producing fruit and livestock feed. This cactus is native to Mexico, from where it had been introduced soon after AD 1492 to Europe, Africa and Asia. The Dutch East India Company VOC probably introduced it in the southwestern part of the Cape Province. By 1750, after the plant had spread to the much drier eastern part of the province, the species had developed into a pest (although it also provided food for less affluent sections of the population).

In 1811, another species was introduced from the Mediterranean region: **oleander *Nerium oleanderi***. After that, the number of introduced species exploded: **tobacco** *Nicotiana tabacum*, **bugweed *Solanum mauritianum*** and various species of ***Acacia, Hakea*** and ***Opuntia***, as well as **water hyacinth *Eichhornia crassipes*** and *Salvinia* (a floating fern) and other species of plants.

The plants were actively introduced and distributed for various ends, such as the production of food and feed; ornamentation of gardens and landscapes (which were found too barren); and extraction of bio-chemicals such as red dyes (from a louse living on *Opuntia*) and tanning substances (from *Acacia* bark). Furthermore, they were used to control sand drift; as cattle fences; to prevent fires; and, last but not least, for the production of wood. Water hyacinth was imported as an ornamental.

Most species were native to Australia and South America, though some were obtained from botanical gardens such as Kew Gardens in London (see also Box 74). Botanical gardens propagated the sale and distribution of seed, using it as a source of income. Missionaries also interpreted their duties quite broadly. Many local invasions started at missionary posts. Later, the forestry commission and private enterprises became the most active distributors of exotics.

Damage and control
Currently, over 50 species are considered to be a small- or large-scale pest. The types of damage caused by these species are diverse. In water:
- Disruption of aquatic ecosystems, irrigation works and hydroelectric installations.
- Impediment of outdoor sporting activities such as angling, swimming and water-skiing.
- Disruption of drinking water facilities; clogging of water intakes and pumping installations, and water evaporation.
- Depletion of groundwater reservoirs.
- Disruption of natural watercourses used by indigenous species.

Since 1905, a series of laws to combat invasive species have been introduced. Various programmes have been launched, the costs of which were not shunned. Annually, $25 million is spent just to combat the water hyacinth, and another $50 million to fight exotic tree species. Making a virtue of necessity, the effort to combat invasive species has led to the creation of unemployment relief works, providing jobs to 20,000 people. In addition, groups of volunteers cut down trees or kill them by girdling. The wood is used as timber or firewood.

Next to invasive plants, South Africa's biotas are being plagued by invasive animals. Particularly harmful is the **Argentine ant *Linepithema humile***, which predates on pollinating insects. About 1000 of the native plants rely on local ants to bury their seeds. The invader does not supply that service and instead dislodges the native ants.

* Van Wilgen *et al.* (2002) give a figure of 21,137 species of vascular plants for all of South Africa, 80% of which are endemic.

Sources: Stirton (1987), Henderson (1995 & 2001) and Van Wilgen et al. (2002)

Bugweed, a poisonous South-American invader in South Africa, Australia and New Zealand.

The share of bio-invasions in the global loss of biodiversity

What is the share of bio-invasions in the decline and endangerment of global biodiversity?

Share in the US

One often cited analysis was made by Wilcove *et al.* (1998). They analysed the factors that play a role in the endangerment of 930 species in the US. Their conclusion: 85% of this group is threatened by habitat loss and 50% by exotic species. Thus considered, invasive species are the *second* largest threat to biodiversity, at least in the US.

However, this analysis has been repeated by Gurevitch & Padilla (2004) and their findings have more nuances:

- Most plants and birds under threat from invasive species are also threatened for other reasons. For instance, the population size of shellfish in North America's Great Lakes had started to drop even *before* the introduction of the zebra mussel, due to habitat loss, eutrophication, pesticides and collectors. Similarly, the population size of cichlids in Lake Victoria had started to decline even before the introduction of the Nile perch *Lates nilotica*, due to railway construction, erosion and damage to coastlines. The authors conclude:
  "Exotic species might be a primary cause for decline, a contributing factor for a species already in serious trouble, the final nail in the coffin or merely the bouquet at the funeral".
- A considerable number of effects of invasive species can be attributed to a limited number of introduced species: rats, pigs, goats, cattle, snakes and some plant species.

The authors do not consider livestock to be invasive, since their numbers are usually checked by humans. They conclude it is wise to focus attention not on bio-introductions generally but on specific *high-risk* introductions; and attention should be given not just to bio-invasions but also to other causes of decline in biodiversity.

Share globally

The most important source of knowledge about the global decline in biodiversity is the World Conservation Union (IUCN). Their most recent analysis was published in 2004, based on the Red List issued in 2003. This list contains 18,318 species for which the causes of decline or extinction are known. The analysis concludes that the following factors have the highest share in species decline:

- Habitat loss or degradation for 33% of the listed species.
- Exploitation (hunting, fishing and poisoning) for 7.6%.
- Exotic species for 6%.

If we follow this list, exotic species rank *third* among the most important causes of global biodiversity loss.

Endangered species are not evenly distributed over different biomes. The species that have become extinct in modern times can be broken down into:

- 570 from terrestrial habitats;
- 222 from freshwater;
- 21 from saltwater (mainly sea birds).

Of the latter group, not a single extinction is attributed to exotic species. That corresponds with the analysis by Wolff (2000) of the Dutch Wadden Sea extinctions. Apparently, marine flora and fauna are not very vulnerable to bio-invasions. Therefore, concerns should focus on terrestrial and freshwater habitats.

What about the future? Sala *et al.* (2000) used model studies to predict which factors will be the most harmful to biodiversity in the next century. Their focus was on terrestrial and freshwater biomes. The result: bio-invasions rank *fourth* after changes in land use, climate change and nitrogen deposition. However, bio-invasions will be dominant in two biomes: lakes and Mediterranean biomes. Islands and rivers, too, are vulnerable to bio-invasions.

Analyses like those mentioned above are highly dependent on definitions applied. For instance, if we focus on diversity at the species level, terrestrial habitats, particularly tropical rainforests, will dominate the scene, since the most species-rich group (arthropods) is well represented there. However, if we focus on diversity at higher taxonomic levels, i.e. genera, families, orders or phyla, other habitats become more important. For example, savannas often have more reptiles than have rain forests. Likewise, seas have more phyla than have terrestrial habitats.

Interactions

It should be emphasized that the above-mentioned key factors are not operating independently. There are several interrelations and feedbacks. For example:

- Climate change enhances bio-invasions, and often habitat loss.
- Habitat destruction or degradation enhances bio-invasions by creating disturbed environments.
- Nitrogen deposition can pave the way for high-resource invasive plants that replace indigenous species.
- Conversely, bio-invasions may increase nitrogen input, particularly invasions of nitrogen-fixing plants in regions lacking such plants.
- Bio-invasions may also lead to habitat loss, as has happened on many islands following introduction of livestock or rabbits.
- Bio-invasions may affect the climate, at least locally, but perhaps even globally (see Box 43).

Leaving apart this latter extreme and still somewhat speculative effect, there are five positive feedbacks that will *accelerate* the process of biodiversity decline.

Next to this decline, the biotas of the planet will be heavily *destabilized* in the next century by climate change and habitat loss as well as bio-invasions.

Invasive species transforming habitats and landscapes

Sometimes an invasive species can cause radical changes in biocoenoses or even in ecosystems, by affecting soil structure, vegetation structure, resource availability and trophic structure.

Invasive herbivores can cause **soil erosion** through over-grazing. Examples: the goat, sheep, cow, pig, deer and rabbit. Trees, too, can cause erosion. For example, the mountains of Tahiti are being overgrown by Miconia *Miconia calvescens*, introduced from South America as an ornamental tree. Its superficial rooting system contributes to landslides. An aquatic example is the snail *Littorina littorea*, introduced from Europe to America. By grazing algae and fixing sediment, it has managed to change mudflats in New England into rocky coast.

By contrast, *Spartina* grass (native to Britain) is causing **sea-floor fixation** in the tidal zone in many parts of the world, providing opportunities for vegetation to settle but reducing opportunities for waders (see also Box 29).

Invasive animals can **root up the soil**. For example: feral pigs on Hawaii and in the Great Smoky Mountains in the eastern US. Various species of plants benefit from this, particularly exotics. Another example: in Ireland's forests, trampling by introduced Reeves' muntjac provides a sprout bed for another invasive species, rhodondendron, which then displaces indigenous species of holly *Ilex*.

The introduction of nitrogen-fixing species in a region where such species do not occur can lead to **nitrogen enrichment of the soil**. This, in turn, can lead to radical changes in vegetation composition and structure. For instance: *Acacia* species introduced from Australia in the *fynbos* of South Africa's Cape Province have raised nitrogen input next to upsetting the scrubland vegetation by introducing a new element (trees), and contributing to groundwater depletion. Another example: the nitrogen-

Spartina grass can fix mudflats, reducing opportunities for waders.

fixing shrub *Myrica faya*, native to the Azores, has settled on Hawaii and spread widely across the lava slopes, where it forms monospecific vegetation. It is also spreading into Hawaii's forests. The nitrogen enrichment is threatening the indigenous vegetation.

Swamping invasive plants and animals (see Box 29) can cause a **radical change in habitat structure**, for instance by hindering germination of indigenous seeds, or by occupying all the space needed by other species, like the zebra mussel is doing in many waters of North America.

Invasive herbivores can **obstruct the regeneration of forests** by their grazing. Examples: the goat and rabbit. Invasive climbing plants, too, can obstruct tree reproduction and thereby kill off entire forests. For example, the forest on Roosevelt Island in the Potomac River in the eastern US, is under threat from the Japanese honeysuckle *Lonicera japonica* and common ivy *Hedera helix*, both native to Europe. Likewise, the isopod *Sphaeroma terebransi*, introduced in the 19th century with wooden ships from the Pacific Ocean, has settled in the mangrove forests of the Western Atlantic. It attacks the terminal roots of trees and thereby hinders mangrove expansion into the sea.

Conversely, invasive pathogens of herbivores can **stimulate regeneration of forests**. In some European countries, for example, the RHD virus recently killed so many rabbits that at some places forest has taken over. In Africa, the rinderpest virus killed so many cattle in the 19th century that much grassland developed into *Acacia* savannas. The famous savanna landscape that exists today is partially a product of a virus!

Much earlier, in 14th century Europe, the plague bacterium killed so many people that much farmland was abandoned and forests could expand. Some of those areas, called *Wüstung* in Germany, have even remained uninhabited until today. Similar events have taken place in America in the 16th century after introduced Old World diseases had decimated Amerindian populations. If HIV/AIDS continues to destroy the farming community in Southern Africa, many a *Wüstung* may appear there as well.

Invasive grasses can radically increase the frequency of **fires**, with significant consequences for vegetation and fauna. Examples are: invasive trees in South Africa's *fynbos* (see Box 40) and cheat grass in the US (see Box 29).

Some plants can **deplete groundwater reservoirs**. For example: the deep-rooted tamarisk *Tamarix ramosissima*, native to the Mediterranean region, has dried up oases and even some rivers in the southwestern US. Fortunately, that effect proved to be reversible when the trees were felled. The water hyacinth can evaporate so much water that **lakes shrink**. An example is Lake Dian Chi in southern China, where 38 of the 68 fish species have disappeared.

Bio-invasions can also **change the climate** locally, and in extreme cases perhaps even globally (see next Box).

Sources: Williamson (1997), Bright (1998), ISSG (undated) and Flinkenflögel & Bol (1992)

Can a bio-invasion change the climate?

If bio-invasions can change habitats and ecosystems, it seems not unlikely that they also can affect the **local climate**. One species that can have that impact is the water hyacinth *Eichhornia crassipes*. This plant evaporates much water, and thus can make a lake shrink and make the local climate more arid. This has happened, for example, in lake Dian Chi in Southern China (Bright 1998).

But can bio-invasions also change the **global climate**? At first sight this seems absurd, but there are indications that such effects have occurred in history. For example, the Little Ice Age (AD 1300-1850) was presumably caused by a significant decrease in atmospheric CO_2 concentrations, as reconstructed from analyses of ice cores and preserved plant leaves. This decrease has partially been attributed to the second plague pandemic starting in AD 1346 and dropping off after 1680 (see Box 5). In Europe, the Black Death (1346-1352) alone killed between 25 and 60 percent of the population. As a consequence, large areas of farmland and rural villages were abandoned, followed by forest re-growth. This caused massive sequestration of the greenhouse gas CO_2, lowering atmospheric CO_2 concentrations. Reoccupation of farms was retarded in some regions by repeated, though less virulent plague invasions during 1400-1720.

Similar effects probably occurred in the Americas after 1492. Invasions of smallpox, measles, mumps, influenza and malaria decimated Amerindian populations, again leading to widespread farm abandonment, forest re-growth and CO_2 sequestration.

Summing up, the assumed causative chain of causes and effects is:

pathogen invasion > less people > less agriculture > more forest > net CO_2 sequestration > lower atmospheric CO_2 concentrations > cooler climate

Although this chain can be regarded as a fact, the quantitative aspect remains uncertain: was the carbon sequestration massive enough to account for the decrease in CO_2 concentrations that took place during the Little Ice Age?

Be that as it may, pathogen invasions can at least in theory affect the climate by drastically increasing tree biomass through eradicating land-cultivating humans or grazing herbivores.

Other mechanisms are thinkable. For example, a severe pandemic of influenza could for a year or more drastically reduce air traffic and thereby the number of sunlight-reflecting contrails in the atmosphere. That in turn would affect daytime temperatures in the lower atmosphere for the time being (as became clear in the days following the September 11th 2001 attacks in the US, when almost all airplanes were kept on the ground). A long-term effect might be less CO_2 emissions and thereby somewhat *less* climate warming.

But there may be still other mechanisms by which bio-invasions can affect the climate.

Sources:
Ruddiman (2003), Van Hoof et al. (2006)
http://www.bbc.co.uk/sn/tvradio/programmes/horizon/dimming

Knock-on effects: how an introduced shrimp decimated the bald eagle – and ecotourism

Bio-invasions can cause knock-on effects: the introduced species cuts the number of another species, which then causes another species to decrease, or increase, etc. This effect has often been intentionally used in agriculture. An introduced natural enemy (species A) of a pest or weed species reduces the numbers of that species (species B), followed by increasing crop yields (species C).

As mentioned in Boxes 32 and 33, introductions may also have unexpected adverse knock-on effects. A dramatic case was the introduction of the freshwater **opossum shrimp *Mysis relicta* in Flathead Lake, Montana, US**. From 1949, this shrimp was successfully introduced in a hundred lakes in the US, boosting salmon production. Between 1968 and 1975 it was introduced in the Flathead catchment, from where it arrived in Flathead Lake in 1981. Here the actual result was quite different. The shrimp displaced several zooplankton species, including the staple food of the salmon. The salmon hardly fed on the introduced shrimp itself and its numbers fell dramatically in 1986.

That in turn caused the collapse of its predators, the bald eagle *Haliaeetus leucocephalus*, which mainly appeared in autumn. The bald eagle fell from several hundreds in 1981 to a mere 25 in 1989. Other predators affected were grizzly bear, coyote, mink and otter. Two gulls, four duck species and a dipper declined as well.

The final victim was tourism, which was mainly related to the bald eagle. The number of tourists fell from 46,500 in 1983 to less than 1000 in 1989.

Source: Spencer, McCelland & Stanford cited in Williamson (1997)

Bald eagles.

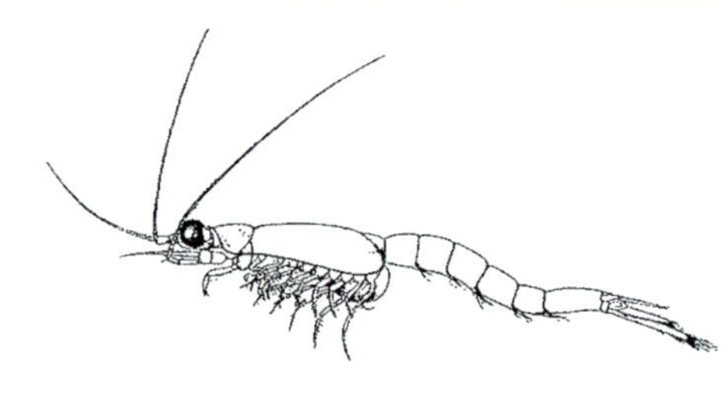

Opossum shrimp (after Williamson 1997).

Biotic resistance to bio-invasions

Humans, animals and even plants have developed natural resistance against invading pathogens and parasites. This system is most highly developed in vertebrates, providing a specific response to a wide range of substances and cells. Since the system has a capacity for memory, it can be activated preventively by vaccination, a practice routinely applied in humans, livestock and pet animals. Wild animals, however, usually cannot be vaccinated, at least not at an acceptable cost.

Invertebrates and plants have a less complicated, but still impressive defence system. For example, plants produce thorns and toxic substances known as phyto-alexins that offer some protection against parasites and herbivores. They also produce signalling substances that attract enemies-of-their-enemies. This is, for example done by the cassava plant (see Box 31). But the protection against enemies offered by these methods is not always effective.

Although biocoenoses lack a defence system, ecologists have used concepts like "ecological resistance" (Elton 1958), "biotic resistance" (Blumenthal 2005) and "resilience". Whilst the defence system of individual animals can distinguish between "self" and "non-self" or "alien" cells, biocoenoses are unable to distinguish between indigenous and non-indigenous species.

However, many biocoenoses have indigenous predators, parasites and pathogens that can attack an invader and reduce its numbers. Furthermore, in many biocoenoses invasive seeds have poor opportunities to germinate. This justifies the term "biotic resistance" (or "ecological resistance" or "biotic resilience") .

Biotic resistance often becomes apparent after it is disrupted: when the system has been disturbed through soil cultivation, soil compaction, intensive grazing, eutrophication, irrigation or pollution. This is the moment at which invasive species often strike (e.g. Lake & Leishman 2004). Disturbed systems can be seen as the "wounds" or points of entry through which invasive species can invade.

This provides us with a first means of reducing the risk of bio-invasions: by limiting disturbance as much as possible. Theories of resource availability and resource fluctuations (see Chapter 3) may provide additional starting points.

One step further is to *reinforce* the resistance to invasions. A frequently applied method is the introduction of an exotic pathogen, parasite or predator to control an invasive pest, often from the native region of the pest or its host. However, such a species is not always available. Furthermore, its effects are often unpredictable, even where a monophagous species is used.

Even more interesting would be the mobilization of *indigenous* parasites, pathogens, predators or herbivores. This type of research is not necessarily a lost case. For instance, it does not seem impossible to activate such species by introducing "crippled" individuals of an invader, analogous to vaccines. If such research has not yet been carried out, it may be worth wile to start with it.

Eradicating invaders

Alien species can be eradicated, at least in theory. The motivation for this will be greater when the species has developed into a pest or is expected to do so.

Islands

Most success stories of eradication are from islands (Genovesi 2005):

- In **New Zealand**, there have been 156 recent successful attempts to eradicate an invasive species.
- Since 1995, various species have been eradicated on 23 islands in **northwestern Mexico**.
- Since 1969, various mammal species have been eradicated on 48 islands near **West Australia**.

These eradications involved mainly vertebrates such as rats, cats and goats.* Some invertebrate and plant invaders, too, have been eradicated, including a fruit fly on the island of Nauru in the Pacific Ocean.

Increasingly smart methods are being used. For example, on island where goats can easily hide, so-called "Judas goats" carrying a transponder are released. These join other goats treating their locality.

Continents

There are few success stories of eradications on continents. The eradication of a polychaete (bristle worm) on an aquaculture farm in California was just a local success. One of the very few large-scale success stories was the eradication of the **malaria mosquito *Anopheles gambiae* in Brazil**. It had been imported in 1929, probably with a rapidly steaming French man-of-war from Dakar, West Africa. The mosquito generated a massive epidemic of malaria. In the 1940s, a large-scale campaign using insecticides did wipe out the mosquito. This could succeed only because this mosquito regularly enters homes and reproduces in pools of water outside the shade of the forest (Elton 1958).

A remarkable success story is the eradication of the **New-World screwworm fly *Cochliomyia hominivo-***

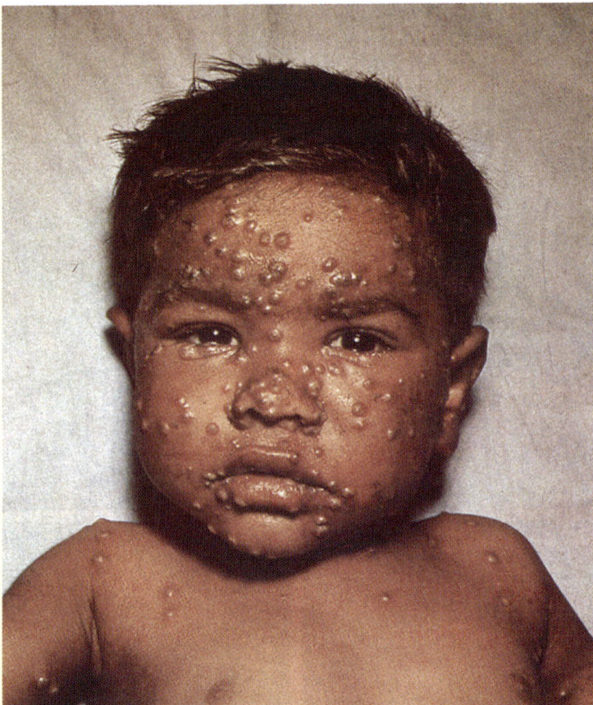

Front of the WHO smallpox recognition card that was widely used from 1971 in endemic countries. Eradication workers searching for cases would show the card and inquire whether anyone had seen a person with a similar rash.

Below: smallpox viruses.

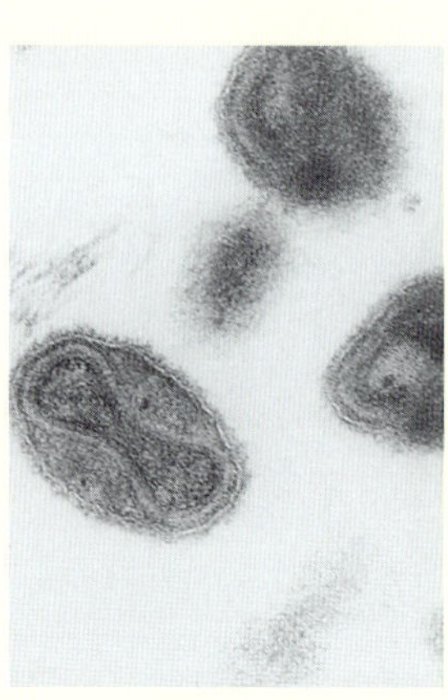

rax. This insect feeds on flesh wounds of warm-blooded animals including humans – hence the second name – where it lays its eggs. Thus it has caused major damage to livestock and humans in the southern US. It was first eradicated in 1959 on the Caribbean island of Curaçao by mass spreading of sterile males. Then the same technique was successfully applied at a much larger scale in the southern US (Groeneveld 2006). Finally it was applied in 1991 in North Africa, where the insect had formed a beachhead. Thus an invasion of a dangerous insect in the Old World was nipped in the bud (Kouba 2004).

Highly contagious animal diseases are being eradicated in more and more countries. Under international veterinary and trade agreements, a country that has eradicated such a disease can obtain significant trade preferences.

Europe

Europe has thus far been fairly reserved in regard to species eradication. At least* 37 programmes have been implemented: 14 in France, 7 on the British Isles, 7 in Spain, 6 in Italy, 2 in Portugal and 1 in Estonia. All programmes were targeted on **mammals**: the brown rat (16x), black rat (9x), rabbit (4x), cat (2x), goat, muskrat, coypu, beaver, Indian porcupine and a species of mink (1x each).

According to Genovesi, there are no documented cases in Europe of eradication of invasive plants and marine organisms. The lion's share (33) of the programmes involves small islands. However, three *local* invaders (muskrat, coypu, and Indian porcupine) have been eradicated in Great Britain, as well as one on the French mainland (a species of mink). Many more opportunities for eradication are available, particularly on small and medium-sized islands and in lakes. The Netherlands is attempting to eradicate the coypu (nutria).

Several European countries are targeting the **ruddy duck *Oxyura jamaicensis***, native to North America, which is crossbreeding with the endangered indigenous white-headed duck *Oxyura leucocephala*.

Global

Eradicating such species that have already spread widely is very difficult, costly, or even impossible. Nevertheless, there is one global success story: the eradication of the **smallpox virus** was completed in 1977. This in fact was the first global eradication campaign – though it was proposed by Edward Jenner as early as 1798. The success was achieved by a global 10-year programme of the WHO based on mass vaccination, supplemented with isolation of remaining infected individuals and on active case-finding, supported by local populations. This spectacular success was possible only because the virus has no reservoir in animals. After eradication, vaccination was no longer needed and was stopped in all but four countries. The virus only survives in two freezers, in the US and Russia, in case it will ever be needed again.

Other likely candidates for a global eradication programme include the poliovirus, the Guinea worm, the hepatitis A and B viruses as well as the rinderpest virus.

* "At least" because not all eradications have been documented. This applies particularly to invasions nipped in the bud.

Medical measures against invasive pathogens

Once a pathogen has spread in a country, protective measures are taken. As for human pathogens, a wide range of methods is available.

The oldest method of all is: **fleeing**. Boccaccio's *Decameron* describes the successful flight of ten Florentines in the plague year of 1348. In many cases, however, fleeing has aided the spread of the pathogen.

Also very old is **quarantine**, particularly the quarantining of ships, which dates back to as early as the Middle Ages. The term is derived from the 40 days chosen for the period based on a biblical source – a lucky choice, in hindsight.

Also of considerable age is the **cordon sanitair**, closing cities or regions off from the rest of the world in order to prevent the epidemic from spreading.

Decontamination is another old technique. A modern example: it is considered important to kill any insects that get inside planes (Spielman & D'Antonio 2001, Bol 2002).

The Netherlands

The Netherlands has a relatively highly developed defence system against invasions of pathogens. First, like many countries, it has various agencies of which the mission includes preventing pathogen invasions, including public health services; medical, veterinary and phytosanitary inspection services; and a Food Safety Authority.

During the previous century, the incidence of **tuberculosis** (TB) decreased to less than 1500 annually, due in part to tuberculosis services as well as control of cattle TB (*Mycobacterium bovis* being nearly as pathogenic as *M. tuberculosis*). However, after bottoming out in the 1970s and 1980s, the incidence of TB is now on the rise again, due to immigration and an ageing population. Some elderly people suffer a reactivation of dormant infection; this is often counted as a new case, though it is in fact recrudescence of TB infection at a young age. Most feared are invasions of multi-resistant TB bacteria.

Methicillin-resistant *Staphylococcus aureus* **(MRSA)** is relatively rare in the Netherlands due to a very restrictive antibiotics-supply policy. Most cases are import cases and are combated intensively in hospitals. Patients transferred from hospitals in Southern Europe (as far north as Belgium) are always isolated and examined for MRSA infection.

There are extensive scenarios for dealing with patients infected with the extremely virulent **Lassa virus** or **Ebola virus**. A handful of hospitals have been equipped to serve as reception centres with stringent bio-security measures.

Protocols are available for potential invasions of many other dangerous pathogens, including the viruses causing smallpox (in case it ever returns), SARS, severe influenza and polio; and the bacteria causing anthrax and plague (both also potential weapons for bio-terrorists). Most viruses can be controlled with vaccination, most bacteria with antibiotics (as long as the bacteria have not become resistant).

Following September 11th, 2001, the Dutch government ordered the purchase of millions of doses of cowpox vaccine, which are stored at the Netherlands' State Institute for Health and the Environment (RIVM). The RIVM also keeps stockpiles of anti-sera or vaccines against other pathogens that may be re-introduced some day, such as diphtheria and plague.

Pollution from the control of invasive plant diseases in Brazil

Bio-invasions of crop and livestock pathogens can lead to structural use of pesticides and veterinary drugs. Brazil provides an extreme example of this. In that country, invasive pathogens not only cause large-scale damage to crops, but are also the reason for nearly the entire use of fungicides and antibiotics.

Fungicides, including banned substances, are used in the cultivation of soy (particularly against soybean rust), sugar cane, coffee, citrus fruit, potatoes (particularly the *Bintje* variety introduced from the Netherlands) and tomato. The latter two crops were commonly sprayed daily, a practice that has been banned. However, spraying is still often combined with irrigation – so-called "chemigation".

Not surprisingly, these practices have caused large-scale pollution of waterways, violation of food residue norms and disease among farm workers. Integrated pest control is still in its early stage.

Source: Lobo Junior (2002)

Last–resort methods to save island birds

In the mid-1970s, the **Mauritian kestrel** *Falco punctatus* had nearly been driven to extinction as a result of habitat loss (through tree felling and the introduction of deer); use of pesticides; loss of genetic variation; and introductions of predators, including monkeys, rats and a mongoose.

An ardent nature conservationist decided to take action and managed to save the species, applying a number of unusual measures:

- Stimulation of egg production, through additional feeding and other means.
- Removal of eggs from nests to be hatched out by incubators and a European kestrel.
- Selection.
- Breeding using artificial insemination.

The kestrels used to breed in trees, were their nests were robbed by monkeys. Efforts focused on the few birds that breed on rocks. Although the danger of extinction has now subsided, the species remains under threat from the mongoose. Nevertheless, the wild population has grown from 4 individuals in 1974 to near 1000 individuals in 2005. Its principal habitat, a mountain forest, was recently declared a nature reserve.

The endemic **Campbell Island teal** *Anas nesiotis* of Campbell Island, some 700 km south of New Zealand, is a non-flying nocturnal bird. Around AD 1900, it was considered extinct. Its rediscovery in the 1970s, in tiny numbers on a nearby rat-free rock, came as a big surprise.

Birds were caught and bred in captivity elsewhere. Subsequently, poisoned bait was used successfully to free the island of the introduced brown rat that occurred there in high densities and had deprived the island of many ground-breeding birds. In 2004 the teal was reintroduced. Only time will tell whether or not it will survive.

Sources: Quammen (1996), World Birdwatch (2003 & 2005) and Gummer (2006)

The Mauritian kestrel was saved from its invasive predators (From: Haliday 1978).

Impacts on the economy and public health

Bio-introductions may have a major impact not only on nature and environmental quality but also on various sectors of the economy and on public health. The impact may be beneficial, detrimental, or both. The sectors most affected are agriculture; livestock husbandry; fishery and aquaculture; commercial shipping; waterways; power stations; and houses and other buildings.[57]

Agriculture and forestry

In most parts of the world, agriculture is largely based on introduced crops. The result has been a strong overall growth in global agricultural production. Because introduced crops are freed from at least some of their pathogens and parasites, they often perform even better than they do in their native regions.

Deliberate introductions can, obviously, be highly beneficial to society and economy. That effect is partly permanent, but also partly negated as a result of unplanned follow-up invasions by pathogens and pests targeting the introduced crop species, causing losses of crop yields and/or control costs. In some cases, however, the negation effect is, in turn, reduced by a tertiary invasion by a natural enemy of the pest species, e.g. a parasitoid wasp. Such counter-invasions are often planned.

Exotic species are a major source of damage to crops. Worldwide, the damage done to agriculture by invasive species is estimated at between 55 and 248 billion dollars per year (Bright 1998). Coherent country-specific studies into the economic costs have only recently started to become available.[58] Pimentel *et al.* published a pioneer survey for the US in 1999 and updated it in 2002 and again in 2005 (Table 4). Their figures, which include the human diseases HIV/AIDS, influenza and syphilis, though not the associated productivity losses, add up to the huge sum of $120 billion dollars annually.

Several authors in Pimentel (ed) (2002) provided figures for agriculture and forestry in six different countries: the US, the UK, Australia, South Africa, India and Brazil (see Table 5). Surprisingly, the highest losses occur in India, followed by the US and Brazil. The total for India is $91 billion. The total amount for the six countries is an estimated $229 billion per year – a low estimate, as no figures are available for several items in various countries

The blessings of biological globalisation: exchange of crops.
Above right: from Old World to New World: wheat (left) and soybean (right).
Below from left to right: from New World to Old World: cassave, potato and maize.

Estimated costs associated with some non-indigenous species introductions in the United States (x $million per year)

Category	Non-Indigenous species	Losses and damages	Control costs	Total costs	%
Plants	25,000			22,658	19%
Purple loosestrife				45	
Aquatic weeds		10	100	110	
Melaleuca tree		NA	3-6	3-6	
Crop weeds		12,000	3000	15,000	
Weeds in pastures		1000	5000	6000	
Weeds in lawns, gardens, golf courses		NA	1500	1500	
Mammals	20			37,477.5	31%
Wild horses and burros		5	NA	5	
Feral Pigs		800	0.5	800.5	
Mongooses		50	NA	50	
Rats		19,000	NA	19,000	
Cats		17,000	NA	17,000	
Dogs		620	NA	620	
Birds	97			1900	2%
Pigeons		1100	NA	1100	
Starlings		800	NA	800	
Reptiles & Amphibians	53				<0.1%
Brown tree snake		1	11	12	
Fish	138	5400	NA	5400	4%
Arthropods	4,500			13,155	11%
Imported fire ant		600	400	1000	
Formosan termite		1000	NA	1000	
Green crab		44	NA	44	
Gypsy moth		NA	11	11	
Crop pests		7000	500	7500	
Pests in lawns, gardens, golf courses		NA	1500	1500	
Forest pests		2100	NA	2100	
Molluscs	88			2205	2%
Zebra and quagga mussels				1000	
Asian clam		1000	NA	1000	
Shipworm		205	NA	205	
Microbes	20,000			15,800	13%
Crop plant pathogens		11,000	600	11,600	
Plant pathogens in lawns, gardens, golf courses		NA	2000	2000	
Forest plant pathogens		2100	NA	2100	
Dutch elm disease		NA	100	100	
Livestock diseases		14,000	NA	14,000	12%
Human diseases		NA	7500	7500	6%
Total				**$120,105**	**100%**

NA = Data not available

Source: Pimentel et al. (2005)

Economic losses to introduced pests in crops, pastures, and forests in the United States, United Kingdom, Australia, South Africa, India and Brazil (x $billion per year).

Introduced pest	US	UK	AUS	SA	India	Brazil	Total
Weeds							
Crops	27.9	1.4	1.8	1.5	37.8	17.0*	87.4
Pastures	6.0	NA	0.6	NA	0.92	NA	7.52
Vertebrates							
Crops	1.0	1.2	0.2	NA	NA	NA	2.4
Arthropods							
Crops	15.9	0.96	0.94	1.0	16.8	8.5	44.1
Forest	2.1	NA	NA	NA	NA	NA	2.1
Plant Pathogens							
Crops	23.5	2.0	2.7	1.8	35.5	17.1	82.6
Forest	2.1	NA	NA	NA	NA	NA	2.1
Total	**78.5**	**5.56**	**3.24**	**4.3**	**91.02**	**42.6**	**228.72**

** Pasture losses included in crop losses*
NA = data not available

Source: Pimentel (ed) (2002)

except for the US.[59] And even for the US there are many "NAs" in the basic figures given in Table 4.

Livestock husbandry

Animal husbandry, too, is based primarily on deliberate introductions, which have often proven highly beneficial to production and the economy. However, some economic damage was caused by animals that had escaped or had been released in the wild. In the US, for instance, the damage done by feral pigs is an estimated $800 million per year, not counting the ecological damage and the harm to public health.

In livestock, follow-up invasions by pathogens can cause massive damage (although such invasions may also reduce damage by feral animals). The main damage is decreased production. In the US, this damage is an estimated $14 billion per year, not counting prevention and control costs (Pimentel *et al.* 2005). In Australia the damage from follow-up invasions is an estimated $249 million, and in South Africa $100 million per year.

Fishery and aquaculture

The main goal of species introductions for fishery or aquaculture is to increase the catch or production. As mentioned earlier, these introductions are often successful, though adverse side effects may appear (see Box 51). Sometimes there are indirect benefits – for instance when a species is introduced to serve as a biological means of pest control.
Pimentel *et al.* (2005) have estimated the damage to US fishery from exotic fish species to total $5.4 billion per year.

Commercial shipping

Bio-invasions to our knowledge offer no benefits to shipping, only costs. These costs are mainly caused by plants such as hydrilla *Hydrilla verticillata*, water hyacinth

Eichhornia crassipes and water lettuce *Pistia stratiotes*, which are clogging waterways. This is also a problem for inland waterways in many warm countries. Aquatic animals and water recreation may also be duped.

In the US, the annual control costs of such plants amount to $100 million. The damage is an estimated $10 million per year. But the share of the shipping industry is probably small.

Waterways

As mentioned above, exotic species can affect waterways and the water supply. Few beneficial effects are known but worth mentioning is the introduction of (sterile individuals of) the grass carp in the Netherlands to fight excessive plant growth in waterways. Since they do not breed in the Netherlands, they are not considered an invasive species.

There are plenty of negative effects on waterways. The three plant species just mentioned as well as *Salvinia* all are able to seriously disrupt water ecosystem services. South Africa provides an example of this (see Box 40).
In the US, a fair share of the $110 million damage and control costs of aquatic weeds results from the choking of waterways.

Power stations and industries

Exotic species offer important benefits to industries and the energy sector, such as agricultural products processed by the agro-industry and energy-generating crops, such as sugar cane in Brazil.

Other invasive species, however, cause major damage, particularly the zebra mussel and other invaders clogging the intake and outlet of cooling water. Both mechanical and chemical methods are used to control this clogging; the latter may cause environmental damage. In the US, the total costs from the Asian clam add up to $1 billion per year, as do the costs of the zebra and quagga mussels.

Houses and other buildings

Species like the dust mite and the common bedbug have accompanied humans for such a long time that it is rather futile to continue categorizing them as invasive species.[60] Heating of houses and other buildings in moderate and cold climates has provided opportunities for new (sub) tropical exotic species like cockroaches. In the second half of the 20th century, the Netherlands and other countries have even experienced something of a microclimate revolution. Many species, including the pharao ant, house cricket, house longhorn beetle and brown dog tick, have benefited (see Box 82). Currently the only available financial figures known to us are for the cockroach in the Netherlands.

Public health

Like all species, pathogens originate in one region and spread from there to other regions, where they can be considered exotic. For instance smallpox, yellow fever, influenza and many other pathogens spread from Europe and Africa to America, where they decimated the Native American population. A host of influenza viruses have spread from East and Southeast Asia to all parts of the world. Recently, the HIV virus (which probably originates in Central Africa) has conquered the world. Syphilis possibly stems from America, while cholera developed on the Indian subcontinent.[61] Malaria tropica probably came from Africa, as did smallpox (Egypt and surrounding regions). The origins of the tuberculosis bacteria and hepatitis viruses are unknown.

Many of these diseases have become endemic in other parts of the world; whether we can still call them invasive species is a matter of definition (see Appendix 1).

The financial costs from a disease consist of the costs of prevention, control and cure plus the indirect costs from productivity losses. The direct costs of disease control can, in principle, easily be expressed in monetary terms; but it is much more difficult to do so for the damage caused. How much is a human life worth, or a year in a human life? How should the costs from sick leave be valued? How do we value the loss in wellbeing? Appendix 2 deals with such questions.

As for the US, UK, Australia, South Africa and Brazil, the total annual costs of fighting HIV/AIDS, influenza and syphilis combined, were estimated by Pimentel *et al.* (2002) at $10.4 billion.[62] Not included are the high productivity losses.

Worldwide, the costs must be considerably higher. For HIV-infection alone, the costs deemed necessary by the WHO control in the middle- and low-income countries alone, are an estimated $9.2 billion. Not included in this sum are productivity losses (see Boxes 64 and 65). For influenza, the total annual global costs, including productivity losses, are an estimated $26 to $60 billion (see Box 66 and Appendix 4).

Influenza pandemics can cause enormous damage. The Spanish flu pandemic of 1918 cost between 20 and 100 million human lives, including 12.5 million in India alone. More recently, the Asian flu of 1957 had over 1 million human fatalities, as did the Hong Kong flu of 1968 (see Box 67). A "normal" influenza pandemic takes 250,000 to 500,000 lives. The costs are on estimated $ 26-60 billion per year (see Box 66).[63] The WHO has repeatedly warned that the current bird flu panzootic is just a

"Environmental" losses to introduced pests in the United States, United Kingdom, Australia, South Africa, India and Brazil (x $billion per year).

Introduced pest	US	UK	AUS	SA	India	Brazil	Total
Plants	0.148	NA	NA	0.095	NA	NA	0.178
Mammals							
Rats	19.000	4.100	1.200	2.700	25.000	4.00	56.400
Other	18.106	1.200	4.655	NA	NA	NA	23.961
Birds	1.100	0.270	NA	NA	NA	NA	1.370
Reptiles & Amphibians	0.006	NA	NA	NA	NA	NA	0.006
Fishes	1.000	NA	NA	NA	NA	NA	1.000
Arthropods	2.137	NA	0.228	NA	NA	NA	2.365
Molluscs	1.305	NA	NA	NA	NA	NA	1.305
Livestock diseases	9.000	NA	0.249	0.100	NA	NA	9.349
Human diseases	6.500	1.000	0.534	0.118	NA	2.33	10.467
Total	**58.299**	**6.570**	**6.866**	**3.013**	**25.000**	**6.733**	**106.481**

Source: Pimentel (ed) (2002)

few mutations away from developing into a severe pandemic. If this happens, the costs might amount to $800 billion, mainly because people will minimise contacts with other people. In addition there are costs from morbidity and mortality. Total costs may amount to 1.250 billion dollars (see Box 67). That is the equivalent of 3% of global GDP.[64]

Environmental costs

Several authors in Pimentel (ed) (2002) have assessed what they call the "environmental" costs of bio-invasions, i.e. the costs not included in the costs for crops, pastures, and forestry mentioned in Table 5. The results for the US, UK, Australia, South Africa, India and Brazil are summarized in Table 6.[65] By far the highest "environmental" costs seem to be suffered by the US, followed by India, but the picture is strongly biased towards the US by the many "NAs" in all other countries.

Total costs

Combining Tables 5 and 6, we get a glimpse of the total costs. For the US, the total amount is estimated at $137 billion, and for India at $116 billion per year. These total figures are underestimates in view of the numerous "NAs" mentioned. For example, no figures are given for the productivity losses from human diseases. For influenza, these costs are much higher than the medical costs. On the other hand the damage attributed to feral cats in the US appears to be arbitrarily high (see Appendix 2).

For the six countries combined, the total estimated cost is **$335 billion annually**. That equals **$240 per capita per year**.

Pimentel *et al.* (2002) even extrapolate this latter figure to the global level in an attempt to give a rough estimate of the total global costs from invasive exotic species: **$1400 billion per year** – equal to **5% of global GDP**. However, this approach is rather crude as long as it remains unclear to what extent the selected group of countries is representative as to costs per capita.

When considering these alarmingly high figures, it should be recalled that some non-native species generate huge economic benefits as well. For the US, Pimentel *et al.* (2005) even assume that exotic species account for 98% of the production of the food system at a value of approximately $800 billion annually. That is 6.7 times the estimated total losses from exotics in the US.

However, the costs and benefits of species introductions are not always evenly distributed among groups in society: often private enterprises benefit from deliberate introductions, while the losses and damages from such introductions are paid for by society. The "polluter pays principle" is rarely applied here (see Box 81). This is one of the reasons why introductions continue at increasing speed.

Costs by taxon

It is interesting to look at these costs from a taxonomic point of view, too. How are the costs distributed among the causative taxonomic groups?

As for the losses in plant production (crops, pastures and forests) in the six selected countries mentioned in Table 5, the plants (weeds) account for the largest share, followed by the plant pathogens (microbes and fungi) and the arthropods. The share of mammals, birds and other taxa is small.

However, the picture drastically changes if we turn to the losses in other sectors given in Table 6. As just mentioned, this picture is strongly biased towards the US, so we must also consider Table 4. Then it becomes clear that, at least in the US, mammals (mainly rats and cats) dominate the scene, followed by microbes causing plant, livestock and human diseases. But again, this picture is in part due to the very high costs attributed to feral cats.

Combining both cost categories makes sense for the US data only (Table 4). There the most costly invasive group appear to be the mammals (32%) and the microbes (combining the costs from pathogens of plants, humans and livestock, adding up to 31%). They are followed by plants (19%) and arthropods (11%). The lower vertebrates (fishes, amphibians and reptiles) are relatively unimportant. Likewise, among the invertebrates, the only group next to the arthropods causing substantial losses is the molluscs (2%). Soil nematodes, too, cause substantial losses of plant production but their costs are included in the microbe figures in an unspecified way.

However, as calculated in Box 66 and Appendix 4, the costs from invasive human viruses are heavily underestimated if we do not include the indirect economic costs from productivity losses. For the US, the total direct plus indirect costs have been estimated by the WHO at $11-18 billion per year, much higher than the $7.5 billion medical costs given by Pimentel *et al.* (2005). That would bring the **microbes** on top of the list of costly invaders in the US.

Potato blight in Ireland and the Netherlands: invasions with dramatic and multiple impacts

The potato was introduced in Europe early in the 16th century, though with little success. Many people thought it could cause disease, whereas others considered it a dreary, vulgar sort of food. By the end of the century, the Irish people were the first to embrace the potato as a highly productive and nutritious crop. Hundred years later it was produced and consumed all over the country and became the Irish people's staple food. That enabled the population to grow from 3.2 million in AD 1754 to 8.2 million in 1845, not counting the 1.75 million that emigrated.

The potato blight disease was observed in the early 1800s without causing major problems. In 1846, however, a virulent variety emerged and quickly destroyed most of Ireland's potato crops. The causative agent was identified as the fungus *Phytophthora infestans*, today classified as an oomycote. The new variety may have snuck into Ireland with a shipment of guano from Chili. In three years' time, 10% of the population died, partly from diseases, to which the population had become more susceptible. These diseases included cholera, which had been introduced in the previous decade. Another 1.5 million people emigrated. In the course of the 1850s, the population had halved to 4 million.

Unlike it is often asserted, food aid was made available to the famished population. Some (British) landowners provided grain. However, the population had lost the ability and material to bake bread. More importantly, the *export* of food continued unabated. Every ship with food entering an Irish harbour saw six ships with food leaving the country. Economy and free trade came first.

In **Belgium** and **the Netherlands**, potato blight had already appeared in the summer of 1845. Between 1/4 and 1/2 of the yields got lost. In the next two years the fungus raged again and the Flanders and Dutch populations suffered famine and became much more susceptible to typhus and cholera. The Dutch population even declined in size. Until today, 1846 even stands as the one and only year since the foundation of the *Bevolkingsregister* (Civil Registration) in 1811, that the Dutch population did not grow but actually shrunk.*

Quite surprising were the political consequences in the Netherlands. In 1847 protesters in the city of Harlingen tried to stop a shipment of potatoes to Britain. In the city of Groningen, unemployed labourers went into the streets demanding bread. The authorities called in the army and five protesters were shot dead. Other towns in the north and east also suffered unrest. When in early 1848 revolutions broke out in several European cities, King Willem II decided that political change was unavoidable and set up the so-called Thorbecke commission. That commission laid the foundations for the Netherlands' system of parliamentary democracy that, albeit with some modifications, still survives. Thanks, in part, to an invasive fungus!

Less beneficial was a third effect: *Phytophthora* continues to be a structural problem in potato cultivation. Recently, a new, more persistent variety of the fungus has been introduced. Even today, the fungus accounts for a considerable proportion of pesticide use as well as its environmental impact.

And that's not the whole story. In the decades following the disasters of 1845/46 less susceptible potato varieties were introduced from the Andes. Again, they did not travel alone and before the mid-1880s another formidable parasite was introduced: potato cyst nematodes of the genus *Globodera*, which attack the roots of the plant. This nematode spreads slowly but proved hard to control and caused increasing damage to potato crops. The environment is affected, too, since the nematodes are controlled in several areas with soil-disinfecting chemicals. *Globodera* has joined *Phytophthora* as a main target for pesticide use in the Netherlands.

* The second time since 1811 the Dutch population is expected to shrink will be within the next decades. Shrinking has already started in a few provinces.

Sources for Phytophthora: Crosby (1972), Bieleman (1992) and Van der Heijden (2001).
Sources for nematodes: J. Bakker (pers. comm.), users.telenet.be/clinckspoor/gewest.htm

Potato field devastated by Phytophthora.

Damage from introduced trees

It is a well-known fact that agriculture and forestry can cause significant damage to nature and the environment. The initial habitat loss is often followed by damage from fertilizer, herbicides and pesticides. Other effects may include groundwater depletion. These effects are associated with indigenous as well as exotic planted species. An exceptionally large amount of damage is caused by some exotic tree species belonging to the genera *Eucalyptus* and *Pinus*, species that dominate the tree plantations in many parts of the world.

Pinus and particularly Eucalyptus plantations are notorious for **depleting ground water**. In eastern Brazil, they have dried up 156 streams and rivers. Around 1990 in Spain and Portugal, farmers staged protests against plantations, with some success, and even destroyed some plantations. In Thailand and Indonesia, farmers have also staged repeated protests.

On many plantations, the **soil is destroyed** by heavy machinery.

Eucalyptus and particularly Pinus species are spontaneously **spreading** from plantations on the Southern Hemisphere and the Mediterranean region, often **displacing indigenous trees**. Of the 2000 introductions of exotic tree species, 135 have led to invasions.

Eucalyptus and Melaleuca trees from Australia are highly flammable (eucalyptus oil!) and cause **forest fires**. Major fires in California in 1923, 1970 and 1992 can be primarily attributed to the planting of Eucalyptus. In Florida, Melaleuca has caused severe fires, even in Miami. A side effect was the heavy, oily smoke that caused traffic accidents and disrupted the electrical supply. The tree survives the fire and even spreads million of seeds over the burnt soil. Seedlings can grow to a height of 2 metres within a year.

In New Zealand, 90,000 hectares of Monterey pines are **sprayed with fungicides**.

This illustrates that even outside their native areas, trees can suffer heavily from pests. Eucalyptus species are often plagued by leaf cutting ants, snout beetles and borers; Pinus species by wood wasps, aphids and nematodes; and Acacia species by leaf cutter ants and bagworms. Furthermore, nearly every plantation suffers from fungi. In Uruguay, Monterey pines were even so heavily damaged by the European shoot moth and an associated fungus that the plantations were abandoned altogether. Today Uruguay is planning a gigantic Eucalyptus plantation.

Source: Bright (1998)

Eucalyptus forests are highly flammable, leading to serious problems in several countries including Portugal (From: www.maos-unidas.pt/.../Incêndios2005).

Diseases of tree plantations and livestock jumping to wild species

Rinderpest by Gustave Doré (19th century).

A common belief is that infectious pathogen and parasites in agriculture, livestock breeding and forestry stem from reservoirs in wild, related species. That has frequently happened. But increasingly they jump in the opposite direction, even to species not closely related. We give some examples from forestry and livestock husbandry where exotic pathogens and/or hosts are involved.

Examples from forestry:

- In **India**, Cercospora needle blight, a fungus that occurs commonly in exotic pine plantations, is attacking two indigenous **pine species**.
- In **Kenya** and **Malawi**, an aphid that began its advance in plantations of Montezuma cypress *Taxodium mucronatum*, has jumped to two indigenous **tree species**, including Malawi's national tree *Widdringtonia nodifolia*.
- In **Australia**, the cinnamon fungus used pine plantations as one of its routes into the indigenous **eucalyptus forests** it killed.

Examples from livestock breeding:

- In 1890, rinderpest was introduced from Asia to the Horn of Africa, probably with livestock transports. That was the start of a dramatic panzootic in **Africa**, accelerated by the seasonal migrations of herds. The herds of the Masai were decimated, depriving their owners from their main food source. Wildlife was also hit hard, starting with the very susceptible wild ungulates including **buffalo, giraffes, gnus, elands, kudus and warthogs**. They were soon followed by their **predators: lions and hyenas**.

- Bovine tuberculosis from cattle has invaded Kruger National Park in **South Africa**, where it is claiming victims among **buffalo, kudus, lions, cheetahs and baboons**.

- In **North America** introduced brucellosis and bovine tuberculosis have infected livestock and then jumped to **bison** and **red deer (wapiti)**. Numerous bison have been shot out of fear that brucellosis-free livestock would be re-infected.

- In **South America**, the decline of **various wild mammals** is thought related to introduced foot-and-mouth disease and cattle rabies.

- Bird flu in **China** and **Southeast Asia** appears to be leaping from poultry to **wild water birds and their predators, including tigers**, as well as vice versa.

- Infectious bursal disease (Gumboro), caused by a virus, can be found all over the world in poultry animals. The virus has even infected two species of **penguin in Antarctica**.

Source: Bright (1998) and Wade (2004)

The insect that almost destroyed the French wine industry

A classic disastrous bio-invasion in European agriculture was the grape phylloxera *Daktulosphaira* (or *Viteus*) *vitifoliae*. This tiny insect feeds like an aphid on fluids from grape roots and leaves, forming galls. Its damage markedly decreases productivity and may kill vines.

Phylloxera was accidentally introduced from the eastern US in the 1860s. The impact of this "great wine blight" was disastrous. It devastated over 2 million hectares of vines in France. Within 25 years, it had all but ruined the wine industries in France, and parts of Germany and Italy. The cultivation could be rescued only by grafting European grapevines onto trunks of resistant American species.

Phylloxera has spread to most vine growing areas of the world, including the western US. Chemicals including endosulfan are being used for control. It has cost California vineyards more than $1 billion damage over the last decade.

Sources:
www.defra.gov.uk/planth/pestnote/grape.htm
ohioline.osu.edu/hyg-fact/2000/2600.html
entomology.ucdavis.edu/faculty/granett/phy_expl.htm

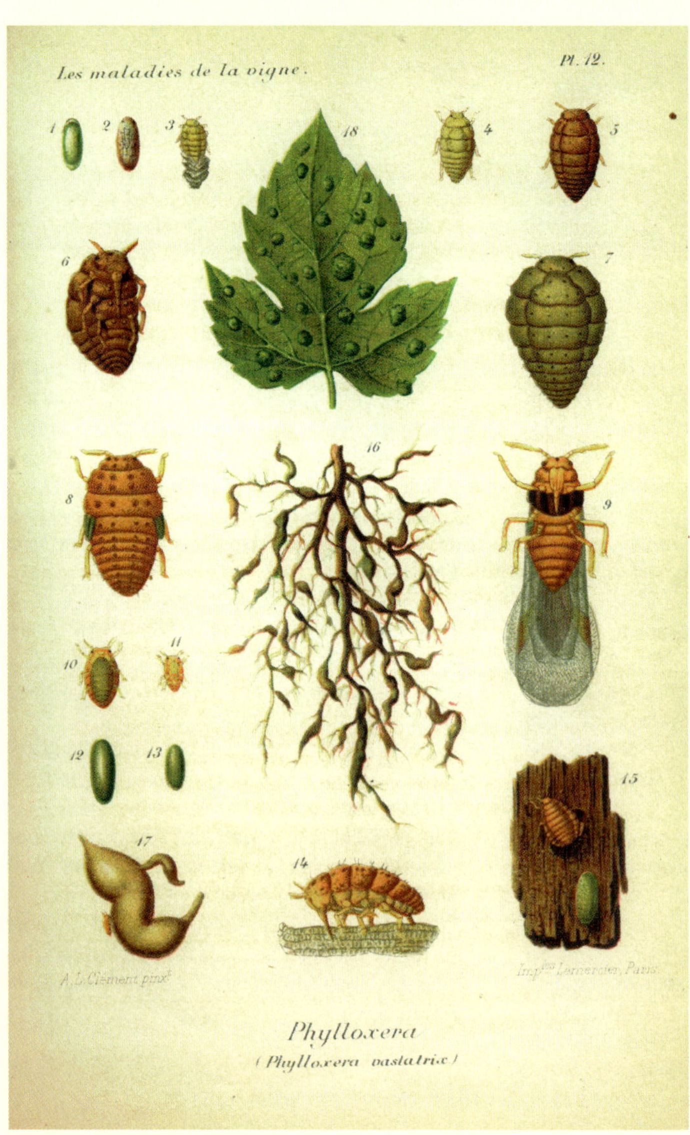

Grape phylloxera
(From: Brunet 1901).

The sweet–potato whitefly: global triumph of a polyphagous, multi–resistant insect

The sweet-potato whitefly *Bemisia tabaci*, also named tobacco or silver-leaf whitefly, is native to Asia. By 1920, it had spread across the world in various crops, including tobacco, melons and tomatoes, without causing much damage.

That changed in the 1980s, when the whitefly began to develop resistance to insecticides. A decade later, it exploded in California and set off on a second triumph across the globe. Underway it developed resistance against a wide range of pesticides, while enlarging its host range. The present biotype B:

- is resistant to most insecticides;
- can attack some 600 plant species;
- can transmit at least 60 species of plant viruses.

In South America, it has made agriculture impossible on 1 million hectares.

This whitefly is currently considered one of the world's most harmful pests.

Source: Bright (1998)

Sweet-potato whitefly *(above)* and its immatures *(below)*.

Bio-invasions in beekeeping

A special form of "livestock-breeding" is beekeeping. It is a relatively small industry in which, however, bio-invasions play an important role.

In 1957, in Brazil, the honeybee *Apis mellifera* (introduced from Europe) was crossbred with the subspecies *A. m. scutellata* (introduced from Africa). The intended result was achieved: honey production soared. But there were serious adverse side effects. The new hybrid form, often called the **Africanized bee**, turned out to be highly aggressive and invasive. It advanced across South America into Central America and reached the US in 1992 (Williamson 1997).

Recently, bees in America and Australia have come under threat from the **small hive beetle *Aethina tumida***. Europe's beekeepers fear an invasion of this species.

In addition, both European and American bees are under serious attack by a mite from East Asia, ***Varroa destructor***.* This mite originally fed only on *Apis cerana*. This Indian honeybee and the western honeybee *Apis mellifera* had long been separated geographically. But because honey production by *Apis mellifera* is more versatile, this species was introduced in East Asia in the previous century. During contact between the two species, *Varroa destructor* jumped to the other species.

In addition, another mite, *Tropilaelaps clarea*, colonized the western honeybees. However, only *Varroa destructor* species managed to spread beyond Asia. It reached European Russia by way of Japan and Siberia. In the late 1970s, it was reported in Eastern European countries and West Germany. The mite has now reached Africa as well as North and South America.

Towards the close of the previous century, the mite reached New Zealand, most likely with an infected bee queen smuggled in. It is now threatening New Zealand's production and export of honey and bee queens, as well as the pollination of clovers and fruit trees, which is valued at $400-900 million annually.

In the Netherlands, the mite was first reported in 1983 and since then has spread across the entire country. The honeybee is essential for the pollination of sweet peppers and tomatoes, as well as for seed cultivation. This is valued at €270 million per year. In addition, the bee pollinates fruit trees and shrubs. Bees are needed by as much as 80% of Dutch crops. Dutch honey actually makes up only 5% of domestic consumption.

The spread of adult Varroa females happens mainly when young bees and drones fly out; when the colony swarms out; through robbery; and by travelling with bee colonies. Without control measures, an infected colony will die within 4 years.

Unfortunately, the mite has become increasingly resistant to pesticides (acaricides). Researchers are now developing biological control with fungi.

* Until recently, all varroa mites were known as *Varroa jacobsoni*. Today, two separate species are recognised. The species that threatens bee colonies is now called *Varroa destructor* (Van Nieukerken, cited by Benedictus 2004).

Sources:

Cook et al. (2002), NCB Bijenziekten Varroa-1
http://home.planet.nl/~gerard.b.w.vos/bijen.htm,
www.ars.usda.gov/is/AR/archive/jun02/form0602.htm
www.bijenhouders-zlto.nl/bijenziekten/varrao-1.htm

Impacts of fish and shellfish introductions

The introduction of fish and shellfish species may or may not have the intended effect on yields. In addition, it can also have other effects.

As far as **yields** are concerned:
Sometimes there is a **persistent increase in yields**, as was the case for the Nile perch in Lake Victoria. Often such an increase is followed by a **relapse**. Sustained increase often requires management. Not uncommonly, **improved yields fail to materialize**. For example: wild salmon. During the past hundred years, various species and subspecies of salmon have been repeatedly moved across the globe, with the intention of boosting local fisheries. In most cases, introduced animals were found to grow less than they do in their native range; there was never any significant increase in yields.

As far as **side effects** are concerned:
In the invaded ecosystem, the **species composition can shift radically**. If the introduction is successful, the new species may start to dominate at the expense of other species. Most likely, this raises the system's vulnerability to ensuing changes, which is also an economic risk. The likelihood and size of such risks are often difficult to predict.

Opinions differ as to the **displacement of local populations**. According to Harache (1992) there are no indications that local populations were displaced. But Pimentel *et al.* (2005) report that in the US, 138 fish species have been imported from elsewhere, and these imported species are threatening 44 indigenous fish species. In South Africa, exotics threaten as much as 60% of endemic species. Bright (1998) mentions the displacement of 10 to 15 indigenous fish species by carps and tilapias in the Mekong River. In a lake in Turkey, the introduced North-American pike *Esox masquinongy* is said to have wiped out three indigenous species of fish. Likewise, in South Africa the rainbow trout is reported to have brought at least one exceptional insect to the brink of extinction: the Gondwana relict damselfly (Bright 1998).

Cultivated species can **escape**, settle in natural ecosystems and cause **disruption** of these systems. For instance: the introduction of the Japanese oyster in the Netherlands (see Box 77), which is increasingly swamping. Fish catches do not always increase after escape because **hybridisation** with local populations can occur.

Perhaps the most far-reaching side effect is that other species, including **parasites and pathogens, hitch a ride** with introduced fish. For instance, 138 species of parasites have been found on carps (Bright 1998). High-risk factors are global oyster transports, which have been taking place for a long time.

There are actually also **positive side effects**. Welcomme (1992) even states that the lion's share of introduced fish species (he gives a long list of these) can be considered positive assets, in the sense that they either have no measurable effect on indigenous fish populations, or make a considerable contribution to the yields of fish farms or fishery in the area. In the Netherlands, the zander (pike-perch) *Stizostedion lucioperca* is a good example.

The transatlantic career of the zebra mussel

The zebra mussel *Dreissena polymorpha* started its global career in the 18th century. The native range of this freshwater bivalve is in the lakes and slowly flowing rivers in the areas surrounding the Caspian Sea and Black Sea (as well as the Aral Sea, but that sea and its surrounding water have become too saline). The animal is highly fertile (each female producing up to 1 million eggs), and can spread very rapidly (the larvae remaining in the plankton for several weeks before settling on hard soil).

Career

The mussel's advance started when canals were constructed in order to connect existing waterways in Eastern Europe. One such canal, completed in AD 1775, connected the Dniepr with the Zapadnyi Bug, enabling the mussel to expand beyond the Black Sea area. More canals followed, and the mussel seized the opportunities. By the early 19th century, it was already found in the Baltic States, and in 1824 it had spread as far west as Great Britain.

The mussel was first observed in the Netherlands in 1826, probably after hitching a ride with wood transports from the Baltic (Kearney & Morton 1970) to Rotterdam. After the inland sea Zuyderzee had been closed off from the North Sea to form the IJsselmeer (literally: Lake IJssel), salinity dropped and the zebra mussel rapidly colonized this new area. By 1938 it could be found throughout the lake (Smit *et al.* 1993). By the 1990s, it lived in all Frisian lakes as well as smaller bodies of water in the southwestern part of the country.

Meanwhile, the mussel had spread north, the first observations in Sweden stemming from 1920. Southward, it has reached the Franco-Spanish border and spread as far as central Italy.

A new milestone in the career of the zebra mussel was reached when it managed to cross the Atlantic, probably in ballast water, or as fouling on ship's hulls. It was first signalled in 1986 in Lake St. Clair near Detroit, and spread quickly into 18 American states and 2 Canadian provinces, using freighters and yachts to move upstream.

Impact

The zebra mussel has a few positive effects. It reduces water turbidity by filter-feeding on phytoplankton, and is a favourite object of underwater photographers.

In the Netherlands, the mussel has greatly benefited two duck species, tufted duck *Aythya fuligula* and greater scaup *A. marila*, of which it is a staple food. Large numbers of these ducks occur in the IJsselmeer. The mussel is also a food source for fish species like roaches, pikes and eels. Water clearing causes a shift to different species, such as water plants and predators relying on eyesight. The water plants provide shelter to prey species from these predators. A negative effect of the mussels is that it clogs water pipes.

Many other negative effects are reported from elsewhere. In France, the mussel has indirectly caused mortality among cyprinid (carp family) fish. The mechanism

Zebra mussel swamping.

Zebra mussels growing on top of each other.

is amazing. The mussel joined another invader from the East, the zander *Stizostedion lucioperca*, introduced for angling, to provide two missing links in the life cycle of a pathogenic trematode, which then was able to attack fishes (Williamson 1997).

However, these effects are peanuts compared to what happened in North America's Great Lakes:

- The zebra mussel lives there in enormous densities of up to 700,000 per m²! It covers and clogs intakes and outlets of power plants and factories.
- Some fish species suffer reduced reproduction because the mussel covers the substrate on which they normally deposit their eggs. These species include some that are favoured on American dinner tables.
- By clearing the water, the mussel has paved the way for invasive water plants.
- It is feared that the mussel will displace indigenous mussel species (North America has 300 species and subspecies, the richest mussel fauna in the world!).
- The mussel accumulates PCBs, dioxins and other toxins present in the silt deposits providing their nourishment. It is feared that birds and other animals feeding on the mussel will get poisoned.

The zebra mussel is combated mainly by mechanical and chemical means, such as chlorine.

Pimentel *et al.* (2005) expect the zebra mussel to invade most freshwater habitats throughout the US. Together with the quagga mussel *D. rostiformis bugensis* (see below) the zebra mussel already causes an estimat-ed $1 billion in damage and control costs annually.

Follow-up invasions

A remarkable side effect of the zebra mussel is that it has paved the way for a follow-up invader from the east: the Caspian mud shrimp *Corophium curvispinum*. It thereby seems to be digging it's own grave. The mussel requires hard substrate to attach itself, using its byssus threads. Hard substrate used to be available in the Rhine, particularly on the groynes. But the Caspian mud shrimp lives in small tubes made of silt, sand and a sticky substance secreted by the animal. Most rocks in the Rhine are now covered with a layer of mud of 1 to 4 cm thick, frustrating attempts by the mussel to attach itself. This has caused a considerable decline of the zebra mussel population in the Rhine since 1989. Recently, the population has recovered to some extent.

In America, a relative of the zebra mussel has recently trodden on the stage and begun its advance: the quagga mussel, a native to the Ukraine. It is currently displacing the zebra mussel locally. In 2006, this species, which according to some experts is conspecific with the zebra mussel, reached Western Europe via the Danube-Main-Rhine route.

Sources: Leppäkoski et al. *(2002), Smit* et al. *(1993), Johnson & Padilla (1996), Williamson (1997), Rajagopal* et al. *(1998), Bright (1998) and Pimentel* et al. *(2005)*

Economic damage from aquatic invaders

Bio-invasions can cause significant damage to fishery and aquaculture. Some invasions cause damage for just a few years, others result in permanent damage and control costs. We give a few examples:

- Damage from the **zebra mussel *Dreissena polymorpha*** in North America to energy plants, industries and fisheries.
  Estimated damage and control costs: $1 billion per year (Pimentel *et al.* 2005).
- Growth of the **Asian clam *Corbicula fluminea*** on cooling water intakes of power plants; the clam also threatens indigenous species through competition. *Damage in the US: $1 billion per year* (Pimentel *et al.* 2005).
- Considerable damage to submerged warf timbers and wooden boats from the **shipworm *Teredo navalis*** in the US, mainly California, since 1990. *Damage: $205 million per year* (Pimentel *et al.* 2005).
- The **European green crab *Carcinus maenas*** has been associated with the demise of the shell-clam industry in New England and Nova Scotia. It also destroys commercial shellfish beds and preys on large numbers of native oysters (Lafferty & Kuris cited in Pimentel *et al.* 2005).
  Damage in the US: $44 million per year (Pimentel *et al.* 2005).
- The (partial) collapse of anchovy and other fisheries in the Black Sea following the introduction of the **ctenophore *Mnemiopsis leidyi*** (see Box 30). *Damage: up to $200 million per year in the late 1980s* (Caddy & Griffits 1990).

- The US has three major invasive **aquatic weeds**: water hyacinth, hydrilla and water lettuce. They are affecting fish and other aquatic animal species, choking waterways, altering nutrient cycles and reducing recreational use.
  Damage and control costs: $110 million per year (Pimentel *et al.* 2005).
- The **water hyacinth *Eichhornia crassipes*** is a global invader from South America. It can cause enormous damage to fisheries, boat traffic and industries depending thereof. In Benin alone, the annual loss of income to fishery and agriculture is estimated at $84 million, a huge sum for a small developing country (De Groote *et al.* 2003). The global costs must be immense.
- The medical costs associated with **toxic species like algae and clinging jellyfish** also carry a hefty price tag. Victims require treatment, prevention campaigns are needed, etc.

The scope of these latter costs cannot yet be accurately determined, but the next example (though *not* referring to an exotic species), can give an indication of what might happen. In Japan, the phytoplankton species *Chattonella antiqua* was responsible for the death in 1972 of large numbers of farmed fish kept in cages (Okaichi 1989). The damage amounted to $500 million. The average annual damage is an estimated $10 million.

Water hyacinth, a South-American invader in the Old Wor

Green crab, a European invader in North America.

Costly invaders:
an ant and a termite in North America

The **red imported fire ant** *Solenopsis invicta* was introduced from South America to Central and North America. It is rampant in Texas, Florida, Georgia and Louisiana, among others.

The ants are a real pest to humans, livestock and nature. They:

- feed on indigenous species of ants;
- negatively impact at least 14 bird species, 13 reptile species, 1 fish species and 2 small mammal species through predation, competition and/or stinging;
- bite humans working in their gardens, damaging their health, sometimes fatally;
- are attracted by electric and magnetic fields, infest computers and equipment, cause short-circuits in homes and in traffic lights, on occasion leading to traffic accidents;
- kill poultry chicks;
- harm agriculture by feeding on soybean and many other crops, digging burrows interfering with combines, and stinging workers.

The species is combated using chemical as well as biological control with fungi, flies and protozoa, but is still extending its range.

The total estimated amount of damage to livestock, nature and public health is an estimated $600 million per year, in spite of $400 million annually spent on control. This adds up to $1 billion per year.

An equally damaging invasive insect is the **Formosan subterranean termite** *Coptotermes formosanus.* It is also causing $1 billion worth of damage in the southern US, particularly in and near New Orleans.

Sources:
Pimentel et al. *(2005)*
http://www.attra.org/attra-pub/fireant.html
http://www.issg.org/database
http://creatures.ifas.ufl.edu/urban/ants/red_imported_fir e_ant.htm

First account:
Spanish Flu in the US 1918

"...Camp Devens is near Boston, and had about 50.000 men, or did before this epidemic broke loose". The flu epidemic hit the camp four weeks earlier, "and had developed so rapidly that the camp is demoralized and all ordinary work is held up till it has passed. All assemblages of soldiers are taboo. ... [patients] rapidly develop the most viscous type of Pneumonia that has ever been seen. Two hours after admission they have the Mahogany spots over the cheek bones and a few hours later you began to see the Cyanosis extending from the ears and spreading all over the face, until it is hard to distinguish the colored man from the white. It is only a matter of a few hours then until death comes and is simply a struggle for air until they suffocate. It is horrible. One can stand to see one, two, or twenty men die, but to see these poor devils dropping like flies gets on your nerves. We have been averaging about 100 deaths a day, and still keeping it up."

"It takes Special trains to carry away the dead... For several days there were not coffins and the bodies piled up something fierce and we used to go down to the morgue...and look at the boys laid out in long rows. It beats any sight they ever had in France after a battle."

Source: "Dr. Roy". Letter dated September 29th 1918, published in the British Medical Journal, December 1979, cited by Kolata (1999)

Bio-invasions catastrophic
to public health

Over the centuries, the human population on a huge continent like Eurasia has developed resistance to many diseases. When these multi-immune individuals and their pathogens come into contact with isolated populations, the result can be disastrous for the latter. To give a few examples:
- Smallpox, measles, mumps, influenza and other diseases decimated the Native Americans.
- The measles killed many Inuit and inhabitants of the Fiji islands.
- Tuberculosis wiped out nearly all inhabitants of Tierra del Fuego.

Ruthless Darwinian selection persisted as long as no suitable vaccines and drugs were available.

The positive side of the current globalisation of infectious diseases is that there will no longer be any "virgin" populations left in the future. This combined with vaccination programmes and other preventive measures, can prevent such dramas as mentioned above from happening again.

However, if a new virulent flu virus appears, we will all be "virgins" and the pandemic could claim tens of millions of victims among the global population. The same may happen if diseases re-emerge against which vaccination has stopped; and even more if vaccines are scarce and the disease spreads more rapidly than such vaccines can be produced.

Yellow fever: a history–maker in the Caribbean region and maritime geopolitics

In the 1640s, the yellow fever virus and its vector, the mosquito *Aedes aegypti*, were introduced in the New World. The first documented outbreak was in 1647 on the Caribbean island of Barbados, the second in 1648 on Hispaniola. The virus was to have a major impact on the history of the Caribbean region.

The mosquito arrived with slave ships from West Africa, possibly ships owned by the Dutch West India Company (see Box 73). The mosquitoes and their eggs survived in barrels of water and the virus in the bodies of slaves or European crew. Slaves were needed for the sugar cane industry that was developing in the Caribbean Islands, starting on Barbados. In Europe, coffee and tea were becoming increasingly popular, and the consumers wished to have their drink sweetened. Whereas before 1600 the consumption of honey and sugar had been minimal, annual sugar consumption in the UK rose to 2.25 kg per person within a century. By 1770 that amount had tripled.

Like many diseases, yellow fever did not hit all groups equally. Africans had a fair degree of immunity, Amerindians and Europeans had not, though those people who had survived the disease were immune for the rest of their lives. This had major consequences for the regional balance of power:

- Epidemics onboard slave ships helped some ships to fall into the hands of slaves.
- British troops besieging Havana, Cuba, in 1762, were hit hard by the virus. The inhabitants of the city had already developed immunity from earlier outbreaks. Out of 15,000 British men, about 3000 sailors and 5000 soldiers died. The British decided to give up and restore Havana to the Spanish in exchange for Florida.
- British troops and refugees carried the disease to Philadelphia. In the 1790s, Philadelphia, then the temporary capital of the young United States, lost one-tenth of its inhabitants to the disease.
- In 1802, Napoleon sent some 25,000 troops to Haiti (part of Hispaniola) to overthrow a rebellion of black patriots. The troops were hit hard by an outbreak of the virus. The commander, before dying himself, asked for more troops, but most of these fell victim too. Of the some 50,000 troops sent by Napoleon, only a few thousand survived to see France again. Yet the revolt was suppressed for some time. But in 1803 a war broke out between Britain and France and in addition the British decided to come to the help of the Haitians. That finally made Napoleon decide to withdraw his troops in late 1803. It was his first major defeat; somewhat simplified: his biological Waterloo. On January 1st, 1804, Haiti claimed independence.

- It was the same epidemic that convinced France to sell Louisiana, another centre of infectious diseases, to the US.
- Yellow fever was a barrier to colonization of French Guyana until 1900, when it was discovered that the disease could be controlled by combating the mosquito.
- After the success of the Suez Canal, opened in 1869, the French (inspired by Ferdinand de Lesseps) wanted to realize an old dream that already dated from the 16th century: digging a canal through Central America. However, the first attempt, starting in 1881, became a complete fiasco because the labourers died like flies from malaria as well as yellow fever. So the classic palindrome "A man, a plan, a canal: Panama!" suggests too rosy a picture. Five years on, the builders admitted defeat and gave up on the project. By then, 30,000 out of 45,000 labourers were in their graves.
- In the early 20th century, Americans resumed the project, with better understanding of the infection cycle of yellow fever, gained by Walter Reed's groundbreaking research on Cuba during the Spanish-American war of 1898. This time the construction of the canal became a success.

If we may draw conclusions from these examples, the yellow fever virus and its vector have influenced the balance of power in the Caribbean region in favour of the Afro-Americans, the Spanish and the US, at the cost of the Amerindians, the French and the British. And in maritime geopolitics it favoured the US at the cost of France.

Sources: Spielman & D'Antonio (2001) and Childs Kohn (2001)

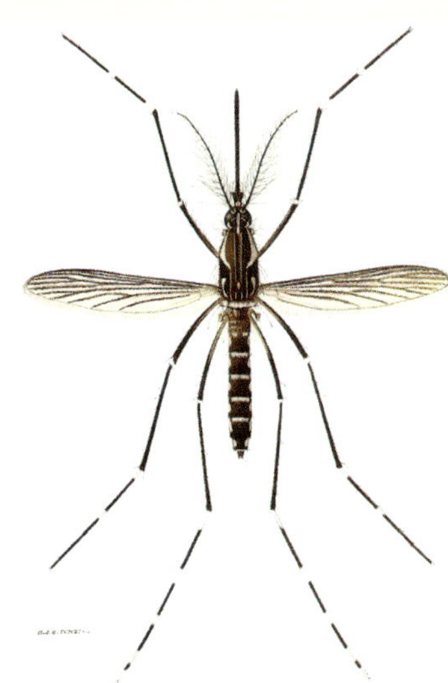

Aedes aegypti, the vector of the yellow fever virus (From: A.J.E. Terzi [from Edwards]).

First account:
AIDS 1981

AIDS is the final stage of HIV infection. There is evidence that the origin of HIV has been a virus that crossed over from primates to humans somewhere in Africa, possibly already hundreds of years ago. Once escaped from its niche it could start spreading globally. The disease was first diagnosed in California, US.

"Pneumocystis Pneumonia — Los Angeles
In the period October 1980-May 1981, 5 young men, all active homosexuals, were treated for biopsy-confirmed *Pneumocystis carinii* pneumonia (PCP) at 3 different hospitals in Los Angeles...

Pneumocystis pneumonia in the United States is almost exclusively limited to severely immunosuppressed patients...

...The fact that these patients were all homosexuals suggests an association between some aspect of a homosexual lifestyle or disease acquired though sexual contact and Pneumocystis pneumonia in this population......[raising] the possibility of a cellular-immune dysfunction related to a common exposure that predisposes individuals to opportunistic infections."

Source: Gottlieb & Weisman (1981)

Contrasting impacts of the HIV virus in South Africa vs. the Netherlands

Like any human pathogen, the HIV virus generates three types of financial costs: direct costs for treatment and for prevention, and indirect economic costs from productivity losses. These costs can be totally different for developing countries compared to developed countries. To illustrate this, we compare the treatment costs in the Netherlands vs. South Africa.

By late 2005, **the Netherlands** counted nearly 12,500 people registered as carrying HIV. That equals less than 0.1% of the population. Most of these are on combination therapy, which is usually successful (see Box 84). Consequently, HIV-related mortality has dropped sharply. Estimates of the medical costs range from €20 to €125 million per year - less than 0.25% of the annual national healthcare budget of €57.5 billion (RIVM 2005). Costs are increasing since more people are being tested; more infections are being detected; combination therapy allows patients to live much longer; and patients have to use the drugs for the rest of their lives. However, the demographic impact of the virus is negligible and the financial impact is very low.

By contrast, **South Africa** had over 5 million people HIV-carriers by 2005. That equals 15-25% of the adult population. For years, mortality has been triple the amount that could have been expected without AIDS. By 2015 the population will even fall 11.5 million short of the number foreseen had the virus been absent.

Combination therapy at current Western prices would cost between €2000 and €10,000 per year for an average patient. For 5 million patients, that adds up to an estimated €10-50 billion annually. Even the low figure would be a huge burden for the Netherlands, but is simply unbearable for a developing nation. Fortunately, the pharmaceutical industry in 2002 yielded to increasing pressure from the civil society and drastically reduced the prices of antiretroviral drugs in developing countries. Present prices in South Africa are around €430 per patient per year, not counting the additional healthcare costs from HIV. That adds up to €2.15 billion per year, still a heavy financial burden for the country.

This burden is exceeded only by the disruptive socio-economic impacts the virus has. AIDS hits society right in the heart. The impact is greatest on the middle group by age as well as income, precisely the group that in every country forms the backbone of society. This brings South Africa on the brink of economic and social collapse. Older people have to get by without support from the younger generation, and instead are forced to take care of their grandchildren. Many orphans are not even being cared for, or have to care for each other, and they have a high risk of contracting HIV themselves.

It has been argued that AIDS could help in the battle on overpopulation. AIDS does actually limit population growth, both globally and regionally. But even in South Africa, the births still considerably outnumber the deaths. The large number of AIDS fatalities in no way encourages family planning. It rather blocks a balanced demographic development.

Global costs from the HIV virus

AIDS was first described in 1981 in San Francisco (see Box 62). Today it is by far the most influential newly emerged disease on the planet. By the end of 2005, there were an estimated 38.6 million people worldwide carrying HIV, including those diagnosed with AIDS; 2.3 million of them were children. More than 25 million people had then already died since 1981 (including 2.8 million during 2005 alone). The total number of cases until 2006 was more than 63.6 million. Since the late 1990s, the incidence is more or less stable; in 2005 it was 4.1 million.

Today, two thirds of AIDS patients live in Africa. Half of them are aged 15-24 years. The continent counts 12 million AIDS-orphans. Originally, the epicenter of the epidemic was Central Africa, but later it moved to Southern Africa.

The costs of the pandemic are enormous and increase rapidly. In the US alone, the healthcare costs for the treatment of HIV/AIDS average $6 billion per year. For the middle- and low-income countries, the WHO claimed in 2005 that $ 9.2 billion dollars was required to control the pandemic. About half of that amount was deemed required in sub-Saharan Africa alone. The sums deemed required in other regions were:

- $ 2.3 billion in South and Southeast Asia
- $ 1.4 billion in Latin America and the Caribbean
- $ 0.89 billion in East Asia and the Pacific
- $ 0.47 billion in Central Asia and Europe
- $ 0.21 billion in North Africa and the Middle East.

In addition to the $ 9.2 billion treatment and prevention costs required, there are productivity losses due to morbidity and mortality. These costs may be even higher, particularly in countries with low rates of unemployment, but may decrease as prevention and treatment become more successful.

Source: www.unaids.org
Source for US: Pimentel et al. (2005)

Global costs from the influenza virus

Almost every year, large areas of the world are affected by one or two new varieties of the influenza virus. These nearly always develop in East or Southeast Asia (see Box 66). During a 'normal' year, on average 10–15% of the global population catch flu, with a case-fatality rate usually below 0.2%.

Some influenza viruses, though, claim many more victims – often when a new combination of a human virus and a bird virus is involved. During the Spanish flu pandemic of 1917–1919 the case-fatality rate was 2.5% in the US; a quarter of the population fell ill and 550,000 people died. Estimates of the global number of fatalities worldwide range from 20 million to as many as 100 million (the margin being so large because there are few reliable data on the death rates in Asia). But even the lowest estimate is higher than the number of soldiers killed during World War I (9.2 million) and World War II (15.9 million). The Spanish flu ranks with the Plague of Justinian (starting in AD 542), the Black Death (1346-1352) and the smallpox and other epidemics that struck the Amerindians in the 16th century, as one of the most disastrous pandemics in history, at least in absolute numbers of victims. But the Spanish virus flu was not a history-maker as was the plague bacterium during the first two pandemics (see Box 5), the smallpox virus in the New World (see Box 18) and the yellow-fever virus in the Caribbean (see Box 62).

Compared to the Spanish flu, the next two "severe" pandemics were mild. In 1957, the Asian flu caused over 1 million deaths, and in 1968, the Hong Kong flu killed almost the same number of people. That is not much higher than in a 'normal' flu pandemic, which claims between 250,000 and 500,000 fatalities in a year.

The economic costs from the flu virus can be broken down into three items:

- Prevention by vaccination. Every year, under the auspices of the WHO, international agreements are made on new influenza vaccines. About 200 million doses of each of the recommended vaccines are produced. Normally, 2 doses at approximately $20–25 per person are given. The total costs thus are an estimated $4–5 billion per year.
- Hospitalisations. For the US alone, these cost are at least $300 million per year. Based on this, we arbitrarily estimate the global costs at $1–2 billion per year.
- Productivity losses.

In Appendix 4, we have made a very rough estimate of productivity losses, based on explicit assumptions and calculations. We estimate these cost to be $21–53 billion per year. Adding this to the above-mentioned estimated costs of vaccines and hospitalisation, **the total global costs from the influenza virus are an estimated $26–60 billion per year.** For the US alone, the WHO has estimated total costs to be $11–18 billion per year. This suggests that the low end of our global estimate may be somewhat conservative.

However, what will happen if the current H5N1 bird flu virus mutates to a lineage that is easily transmitted from human to human, causing a severe pandemic? The World Bank estimates in a 'Severe' scenario (where 35% of the world population is infected and the case-fatality rate is 3%, resulting in 1% mortality) the number of excess deaths could amount to 70 million globally.

As for the economic impact, the indirect costs alone could rise to $800 billion in just one year, or 2% of global GDP. These costs are expected largely because people will minimise contacts with other people, which will disrupt the economy. In addition, there will be costs from morbidity and mortality. **Total global costs from a highly virulent and contagious flu virus may amount to $1250 billion – 3.1% of global GDP.**

Sources: Childs Kohn (2001), Kolata (2001),
M. Brahmbhatt on web.worldbank.org.
Source for number of sick days: WHO
Source for global GDP: IMF World Economic Outlook Data-base, April 2006

Bird flu:
candidate for a new severe pandemic

In 1997 the world was put on alert by cases of bird flu among chickens in Hong Kong, caused by subtype H5N1 of the influenza virus, the fear being that the virus could be transmitted to humans. Until then it had been assumed that pigs are the 'melting pot' in which bird flu viruses can mix with human viruses (a process called re-assortment) to create new viruses threatening humans, as had happened in the pandemics of 1957 and 1968. But then it turned out that mutations generating a virulent strain could occur directly. (Recent research has shown that this shortcut was also used by the H1N1 subtype that caused the devastating Spanish flu pandemic of 1917–19). Of the 18 patients in Hong Kong, 6 died. In all haste, a million and a half chickens were destroyed, which proved successful: there were no new cases. A dangerous zoonosis appeared to have been brought under control.

However, there were some additional minor outbreaks in 2001 and 2002/2003, and since late 2003 we have been confronted with a fast-spreading bird flu epizootic, again of subtype H5N1. The disease started in China and Southeast Asia, where the authorities in some countries (Thailand, Indonesia) initially kept the news under wraps for fear of damaging exports. By March 2006, 200 million poultry animals had died of bird flu or been culled as a precautionary measure. This appeared to be effective in Hong Kong, Japan and South Korea, but less so in continental China, Vietnam, Thailand and Indonesia. China started a massive vaccination campaign. In 2005 the disease spread to Northern and Western Asia, Turkey, and Europe, and even sub-Saharan Africa, and thus became a panzootic. Initially, the virus was mainly spread by the transport of chickens (including fighting cocks) and by movements of poultry workers. Later, it was also spread by migrating birds, particularly ducks and geese. It has been found that apparently healthy birds, including migratory birds, can carry and shed the virus.

Bird flu or avian influenza (AI) is caused by subtypes of the influenza A virus. Based on the surface proteins hemagglutinin (H) and neuraminidase (N), the RNA virus has 16 H subtypes and 9 N subtypes. Each subtype has several sublineages or strains. In wild birds, many of these viruses occur in the intestines without causing disease in

Poultry: source of the next severe pandemic?

the animal, but these low-pathogenic strains can develop into high-pathogenic strains, particularly when they have invaded poultry farms. Such strains are then able to cause disease in poultry as well as in wild birds (mainly water birds).

The virus can be transmitted in various ways, ranging from faeces, saliva and mucous membranes to contaminated cages and water. Cats, tigers, leopards and pigs can contract the disease by eating infected birds or carrion. Most cases in tame birds, particularly the low-pathogenic types, probably go unnoticed, but there are also severe cases leading to multi-organ failure and death within two days among 90% to 100% of infected animals.

High-pathogenic H5N1 was first isolated in 1966 from geese in China's Guandong Province. In subsequent years it was repeatedly isolated from poultry and occasionally caused disease in humans. By December 27th, 2006, the number of fatalities had risen to 157 worldwide, with most cases in Indonesia (57) and Vietnam (42) as well as 4 cases in Turkey, close to Europe. The greatest genetic diversity is still found in southeast China, indicating that this region is the epicentre of the virus and that the virus has become endemic there.

There are two reasons for concern. First, the case-fatality rate is 60%, much higher than the 2.5% in the Spanish flu of 1918 (which, however, took many more victims because of the much higher percentage of cases, e.g. 25% in the US). Second, the virus did not originate from re-assortment in pigs (as in the pandemics of 1957 and 1968) but from mutations (as in the much more devastating pandemic of 1918).

On the other hand, the risk of transmission to humans is still small, and so far almost all patients appear to have been infected through intensive contact with poultry. In the handful of cases where person-to-person transmission is thought to have occurred, the disease was apparently no longer contagious after it had been transmitted to the second patient.

Will this panzootic develop into a pandemic? No one can know for sure. Until now, pandemics of the influenza A virus have been caused only by subtypes H1N1, H1N2, H2N2 and H3N2. As for H5N1, few cases of animal-to-human transmission and even fewer cases of human-to-human transmissions, have been reported. However, a few mutations may suffice to drastically increase the risk. The damage would be greatest if our immune system turns out to be completely naive to the virus and cannot respond effectively.

A recent study estimates that if a virus similar to the one that caused the Spanish flu would emerge, it would kill between 51 and 81 million people within a year, doubling annual global mortality. As in 1918, mortality rates would be unequally distributed, with rather low rates in rich countries and high rates in poor regions. Approximately 96% of the victims would fall in developing countries due to lack of access to vaccines, antiviral drugs and medical care, and high HIV/AIDS prevalence (Murray *et al.* 2006).

The WHO and national healthcare and veterinary services have mustered all hands to monitor the disease and bring it under control. Methods include culling and destroying poultry, vaccination, holding poultry indoors, standstill orders on poultry transport and public information campaigns. However, specific vaccines against the present sublineages of H5N1 do not yet exist. Researchers are working hard to develop such vaccines and clinical trials are underway. But by the time these vaccines are introduced on the market, they may already have lost their efficacy, since the virus changes frequently.

The last line of defence is anti-retroviral drugs, but these may not be effective either, since many virus isolates appear to be resistant against some of the drugs (amantadin and rimantadin). Oseltamivir (better known as Tamiflu) inhibits the virus but does not eliminate it. New drugs are in the pipeline and are currently undergoing broad clinical trials.

One positive difference with 1917–19 is that many patients then did not die from the virus, but from a secondary infection with a pneumococcus. Today we have antibiotics to control such bacterial infections.

The SARS epidemic (caused by a corona virus) in 2002 and early 2003 may have been a blessing in disguise, since it was a general rehearsal that made us extra alert and ready to respond.

Other invasive zoonoses

Animal vectors transmitting zoonoses occur in a wide range of taxonomic groups, mainly insects, mites, birds and mammals.

Anthrax is a cattle disease caused by *Bacillus anthracis* (often its spores). The most dangerous forms of the disease are septicaemia and pneumonia. Anthrax has been eradicated in many countries, although spores may still survive in the soil locally. Of course the disease may be reintroduced accidentally or even intentionally. Powerful sources of infection are diseased mammals and their carcasses. High-risk groups are people who make brushes, or process wool or wool products (such as carpets) from the Middle East.

BSE and **Creutzfeldt–Jakob** also come from cattle. They are caused by a prion: a deformed brain protein that "infects" its environment through a cascading chain reaction, causing more and more proteins to adopt a faulty tertiary structure. It can cause a form of Creutzfeldt-Jakob disease in humans having consumed meat from infected animals (Lieverse 2001). In the past, this ailment used to occur sporadically, mainly in elderly people. The new variant occurs predominantly in young adults, which in the UK led to the detection of the epidemic. In the long term, a maximum of 200 fatalities are expected. Since the prion was exported with meat, cattle or feedstuff, some victims in other European countries were foreseen as well (Bol 2001). In the Netherlands a second victim was diagnosed by June 2006.

Dermacentor reticulatus is a tick feeding on mammals such as cattle, ponies, horses and dogs in the tropics and subtropics, including

"Disinfected" banner used on one of Beijings airports during the SARS epidemic in 2003.

the Mediterranean region. Dogs returning with their owners from holiday can transfer the tick to temperate climates. It was generally assumed that the tick could not survive in the Dutch climate, but recently populations were found on two locations, which may be the start of an invasion. The tick is a vector of the protozoan *Babesia divergens* that causes babesiosis in dogs and horses, which may be fatal. It can also transmit the *Borrelia* bacterium that causes Lyme disease (see below) as well as a rickettsia that may cause Mediterranean spotted fever. These diseases, too, can be fatal to humans. In western Siberia, the tick is also suspected to transmit the virus that causes Omsk hemorrhagic fever from muskrats to humans, particularly muskrat trappers.*

The **Hanta virus** commonly occurs in deer mice in North America, as well as in the bank vole and other rodents in Northern and Western Europe. It is one of the causes of a haemorrhagic syndrome, just like the tropical Lassa and Ebola viruses mentioned in Box 55. In the Netherlands, patients carrying hanta virus have generally contracted it in other countries. Following an incubation period of up to 10 days, the disease progresses to a flu-like state, followed by pneumonia and severe kidney damage, both of which can be fatal.

Influenza viruses often originate in poultry and pig husbandry in China (see Boxes 66 and 67). Hence in many - though not all - cases, influenza is a zoonosis. The virus is airborne and thereby can spread fast from human to human. Airplanes can spread new strains across the globe within a matter of days, particularly if patients already start spreading the virus before getting sick, and if air filters are not timely replaced.

Listeriosis is a zoonosis because the causative bacterium has its reservoirs in ticks, small mammals and deer. It occurs in Eurasia and North America. In Western Europe, the symptoms were described in the 19th century. Back then, however, the cause was not yet known. By now, it has become clear that the spirochete bacterium *Borrelia burgdorferi* is responsible. The name Lyme stems from the American town of Old Lyme (Connecticut), where the cause of the disease was first determined. However, there are actually more species of *Borrelia* in Eurasia than in America, suggesting that the species originates in Eurasia.

Humans contract the disease through a tick bite. In the Netherlands the sheep tick *Ixodes ricinus* transmits the bacterium. Complications include arthritis and meningitis. Although Lyme disease may be indigenous in the Netherlands, the huge increase in tick bites and clinical cases suggests that there is either import of ticks from Southern Europe, Asia and/or America (research into this is currently underway) or import of virulent lineages of *B. burgdorferi*. The increasing popularity of recreation in nature may also contribute, but by itself cannot explain the increase. In 2005, tick bites were responsible for 17,000 patients requiring treatment (Hofhuis *et al.* 2006).

Neurocysticercosis is a human brain disease caused by larvae of the pork tapeworm *Taenia solium*. In the New World it did not occur until Europeans introduced the pig, which is an obligate intermediate host of the parasite. The disease is now a major source of morbidity in Latin America (Bryan 1999).

Psittacosis (ornithosis) is a potentially fatal lung disease caused by the bacterium *Chlamydophila psittaci*. The disease was first reported in Europe in 1876 and was traced back to the import of diseased parrots from the tropics. In 1929-30 there was even a pandemic of approximately 1000 cases, including 200 to 300 deaths in countries importing birds from South America. The bacterium thrives in spaces where birds are kept permanent-

ly. It occurs not only in parrots and para[...] other birds including canaries, pigeons, gull[...] [...]nes.

SARS (Severe Acute Respiratory Syndrome) is a new emerging disease that broke out in China in late 2002 and was identified in early 2003. The virus was spread fast through air traffic across the globe, among others to Canada and some European countries. The pathogen appeared to be a type of corona virus that until then was not considered a danger to the respiratory system. One of the possible reservoir species is the masked palm civet *Paguma larvata*, which is traded at Chinese markets.

The previously mentioned **plague, yellow fever and HIV/AIDS** can also be considered zoonoses. Plague is caused by bacterium-infected fleas on rodents; yellow fever has a sylvatic reservoir among primates (forest yellow fever); and HIV (Human Immunodeficiency Virus) probably has developed from SIV (Simian Immunodeficiency Virus).

Regarding the future, it looks like the significance of zoonoses will increase rather than decrease. No less than three quarters of human emerging diseases are caused by zoonotic pathogens (Taylor *et al.* 2001). Since 1980, more than 35 new infectious diseases have emerged in humans, approximately 1 every 8 months (Karesh *et al.* 2005).

** Source: http://www.phac-aspc.gc.ca/msds-ftss/msds113e.html).*

Box 69

Species introductions with private benefits and public costs

Many plants and animals are imported by private enterprises in order to make a profit. In many cases, this is successful. But as is more often the case: when there is damage, it is often society at large that pays for it. The principle that "the polluter pays" is still rarely applied in these cases.

Some examples:

- In Australia, 466 species of **feed crops** have been introduced. Of those, 21 proved to be profitable to the livestock industry, 60 were invasive, and just 4 were both profitable *and* non-invasive. The livestock industry's profits are minimal ($2 per hectare), but the costs to control invasive species are hefty – $30-120 per hectare. These costs are paid for by the Australian society.

- The recent **Nile perch** industry near Lake Victoria is primarily focused on export; the local population hardly benefits and has been displaced from their traditional fishing and processing work.
- Plantations of **exotic trees** for the pulp industry in South America and Southeast Asia have depleted groundwater reservoirs at the expense of local agriculture and forestry.

Private companies are eager to profit from the "initial peak" in the cultivation of introduced species. But after that peak, yields often decline, either because resources become exhausted or because diseases and pests take their toll.

Source: Bright (1998)

Trade in bush-meat and wild animals: a vehicle for pathogens

Animal diseases including zoonoses can be spread with trade in livestock, meat, wild animals and bush-meat. The cases of foot-and-mouth disease (see Box 20) and bird flu (see Box 66) illustrate the risks involved in trade of livestock and meat. Trade of wild animals and bush-meat may be even more tricky business.

Consumption of bush-meat is an old tradition among local people in tropical forest areas. Trade in wild animals and bush-meat is equally old, but the volumes and the scale are increasing. In some exotic foods stores in New York, bush-meat is today considered a delicacy.

The biodiversity involved in local and international trade is enormous. For example, in one market in Thailand 276 species of birds were counted within half a year. On a market in Guangzhou, China, such widely different animals were found to be traded as ferret badgers, masked palm civets, barking deer, wild boars, hedgehogs, foxes, squirrels, bamboo rats, gerbils, various species of snakes and endangered leopard cats, along with domestic dogs, cats and rabbits.

Among the pathogens found spreading with trade of **wild animals** are:
- H5N1 type A influenza in hawk eagles illegally imported to Belgium from Thailand.
- A paramoxi-virus highly pathogenic for poultry in a shipment of birds from Pakistan to Italy.

- Monkey-pox virus introduced with wild rodents from Ghana and transmitted to humans in the US.
- Chytridiomycosis, a fungal disease considered a major cause in the extinction of 30% of the amphibian species worldwide, spread widely with African clawed frogs.

Export of **bush-meat**, too, has a risk of spreading diseases. Consumption of apes and monkeys in Africa has presumably played a role in the transmission of SIV and Ebola to humans. SIV of various types has been found in 13 primate species in Cameroon alone. As for Ebola, it has been found that humans handling ape carcasses can contract this deadly disease.

It is commonly known that increased consumption, trade and smuggling of bush-meat and wild animals are causing overexploitation of species. In addition, they represent a growing global biohazard to humans, livestock and wildlife alike.

Sources:
Karesh et al. *(2005)*
www.cdc.gov/ncidod/EID/vol8no5/01-0522.htm
http://www.sciencedaily.com/releases/2005/02/050218155913.htm

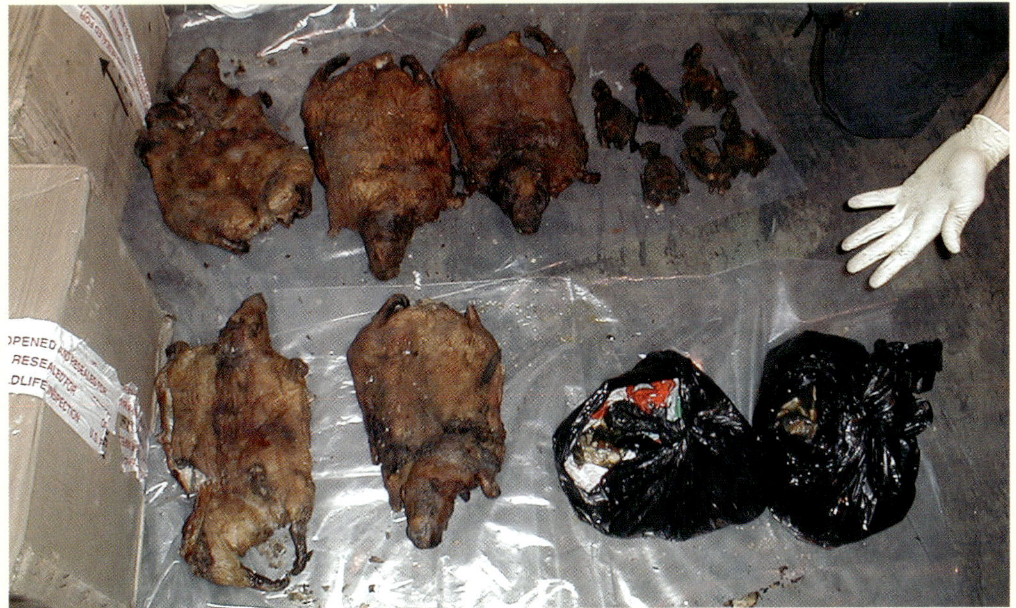

African bush-meat: a possible source of pathogen invasions.

Invasive pathogens as weapons of mass destruction

Many pathogens are potential weapons of mass destruction. Exotic pathogens present a particularly large threat, since the population has not had a chance to build up resistance against these invaders

Wars

In the history of warfare, not many cases are known in which diseases were intentionally spread. Some known (presumed) cases are:

- In 184 BC, Hannibal ordered that pots filled with **serpents** be thrown onto the decks of enemy ships.
- The **plague bacterium** was repeatedly used as a weapon of war, e.g. by hurling plague-infected corpses at the enemy using catapults. A well-known instance occurred at Caffa (now Feodosija, Ukraine) in 1346, when Mongols are thought to have catapulted the corpses of plague-infected individuals into the city.
- **Anthrax** was used as a weapon of war by the Spanish as early as the 16th century in some areas, including the Mediterranean region. They used a mixture of scrap metal, glass and cadavers of anthrax-infected animals as ammunition. The result: horrible wounds and a gruesome death. At that time, anthrax actually already occurred throughout Europe. Another case is the alleged use of anthrax by Japan during World War II.
- **Smallpox** was probably deliberately used by the British against Delaware Indians in the 18th century. British supreme commander Sir Amhurst wrote a report suggesting that contaminated blankets be distributed among Native Americans. Just before, that method had already been applied. The use of fomites was an accepted method to spread the disease.
- During World War I, both parties deliberately spread **horse diseases, mainly glanders**. Horses and mules then still played an important role in combat. It is not known, however, whether the reported diseases are to be attributed to these attempts.

Agriculture, too, may be a target of biological warfare. At least nine countries had documented agricultural bioweapons programmes during some part of the 19th century

Terrorism

For terrorists, too, pathogens are potential weapons of mass destruction. Today, the risk is no longer just theoretical. From suicide attacks using airplanes as bombs it is not a big step to bio-terrorism. Even a single individual can cause massive damage and panic. This became apparent when in the autumn of 2001, an unknown terrorist with unknown motives spread spores of anthrax through the mail in the US.

Some extremist animal-rights activists, too, employ violence, such as setting fire to slaughter-houses. In one extreme case in the Netherlands, a prominent politician running for prime minister in 2002 was killed. In 2001, during the foot-and-mouth crisis in the UK and the Netherlands, an American animal-rights activist threatened to spread foot-and-mouth disease virus on American livestock markets.

The US Centers for Disease Prevention and Control have listed 24 pathogens that could be used for bio-terrorism against **humans**, ranging from the anthrax and plague bacteria to the smallpox and Ebola viruses. Interestingly, the list also includes multi-drug resistant tuberculosis bacteria.

Potential weaponry for use in **livestock** terrorism includes foot-and-mouth disease, classical swine fever (or hog cholera), African swine fever, Newcastle disease, bird flu and rinderpest. Most of these are exotic to the US or have been eradicated. The costs and damage of a foot-and-mouth outbreak alone could be $10-34 billion.

Crop cultivation offers at least eleven possibilities for bioterrorism, including citrus greening, Philippine downy mildew (in maize), soybean rust, plum pox (in stone fruits), brown rot (in potato) and bacterial leaf strike (in rice), all of them exotic or eradicated in the US. Soybean-rust alone could cause damage worth $8 billion per year.

Agricultural bio-terrorism increasingly gets public attention. To illustrate this, a global internet-search in September 2006 yielded more than 1.4 million hits for "livestock terrorism" and even more than 2.6 million hits for "agricultural bio-terrorism". Not surprisingly, the US has made this a political priority. The Senate held hearings and since 2002, the country even has an Agricultural Bioterrorism Protection Act.

Outside the US, there are major threats, too. For example, it might not be impossible for a bio-terrorist to eliminate almost the entire global production of rubber. Rubber production has moved from South America to Thailand, Indonesia and Malaysia. One reason was that the rubber tree *Hevea brasiliensis* had been decimated in its native range by a fungus,

Bioterrorism against agriculture is considered a serious threat in the US (From: Monke 2005).

the South-American leaf blight *Microcyclus ulei*. If this fungus were to be successfully introduced in Southeast Asia, it would cause enormous global economic damage, considering the key role of rubber in our car-based economy.

Many pathogens are not difficult to obtain, and the primary targets (humans, crops and livestock) are easily accessible and impossible to protect. In the case of agroterrorism, the risk for the terrorist is small. By contrast, livestock is becoming more vulnerable because an increasing number of countries have stopped vaccinating against highly contagious diseases. This is heavily stimulated by international veterinary and trade agreements. Thus, more and more "biological powder kegs" are being built up. In biogeographical terms, these countries are attempting to become disease-free "islands" with virgin populations of livestock. Invasion biology and history (particularly the Columbian Exchange) show the biological danger of such a policy. Another danger is that experts are no longer familiar with the disease

and may easily misidentify the pathogen in the crucial early stage of an outbreak.

In the Netherlands, too, the risk of agro-terrorism is not entirely imaginary. The country has a disproportionately large livestock industry. One might speculate that pig farms are most at risk – they may be target for extremist Islamic terrorists (who consider pigs to be impure) as well as extremist animal-rights activists (who seek to put an end to factory farming).

Sources:
Wars: Geisler (1994), Crosby (2000), Davison (2005) and Monke (2005)
Terrorism: Office of the US Secretary of Defence (2001) and Van der Weijden (2001)
www.actionbioscience.org/newfrontiers/davis.html,
www.airpower.maxwell.af.mil/airchronicles/battle/c hp10.html
www.cidrap.umn.edu/cidrap/content/bt/bioprep/biof acts/bioterr-overview.html

The Netherlands and bio-globalisation

The biogeographical position of the Netherlands

Every country has a unique position in geography as well as *bio*geography.

The position of the Netherlands can be characterized as follows.

For terrestrial ecosystems, relevant factors are:
- The Netherlands is part of the huge Eurasian continent, which contains relatively many species and therefore is somewhat less susceptible to bio-invasions than other continents are.
- On the other hand, the country has some very young ecosystems, the polders; the youngest polders are just 50 years old and are still being colonized.
- The Netherlands is a small country, not separated from the European hinterland by any barriers; therefore it has very few endemic species, and there is only a minimal chance that a bio-invasion results in the complete extinction of a species.
- Conversely, an invasive species that has settled in the Netherlands can easily spread across Europe. A current example of this is the Egyptian goose *Alopochen aegyptiacus*.
- The country has large wetlands where millions of migratory waders, ducks and geese winter or stop. Combined with huge concentrations of poultry farming, this makes a global bio-risk, although most poultry are housed in closed systems.

For aquatic ecosystems, the following factors are relevant:
- The Netherlands borders the sea and has the largest seaport outside Asia: Rotterdam. This has resulted in a relatively large number of contacts and opportunities for species interchange with overseas areas. It is no coincidence that in

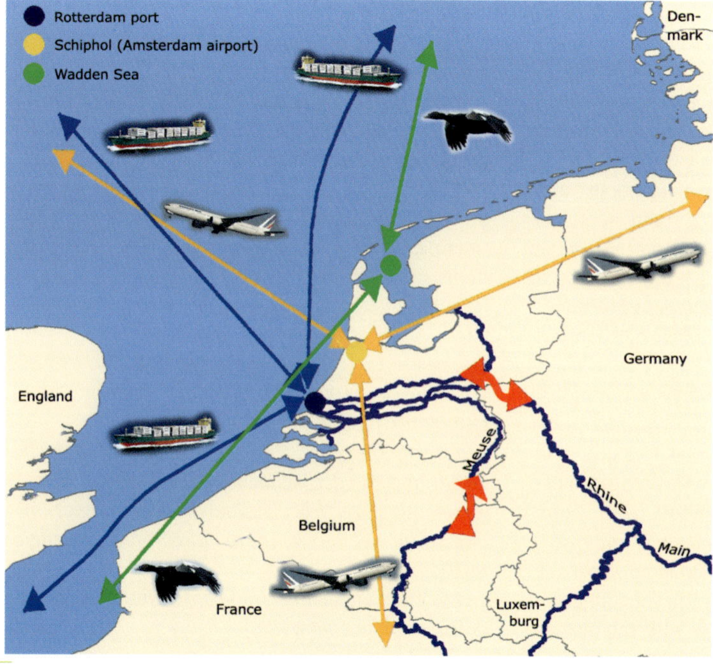

Figure 7.
Map of the Netherlands showing natural and artificial pathways that contribute to bio-invasions

Rotterdam and the surrounding area a large number of exotic plant species are found, to which even a special book has been devoted (Andeweg & Florusse 2002). More surprising is that few exotic animals have been found so far.[66]

- The Netherlands has excellent water connections to the European hinterland: two major rivers (the Rhine and the Meuse) have for millennia flowed through the country on their way to the North Sea, creating a permanent influx of water species, including occasional new species. The influx was intensified since the Rhine-Main-Danube canal was completed in 1992, creating a direct freshwater connection between the Rhine delta and the Danube delta in Romania. Although freshwater organisms can move in both directions, most invasions are in the direction of the Rhine Delta due to the dominant direction of the water flow through the canal. At Basel, exotic species already make up 95% of animal biomass in the Rhine (Wittenberg 2005).
- The Dutch coastal waters are part of the geologically young North Sea and are therefore relatively susceptible to bio-invasions (Wolff 1999).
- The Dutch freshwater systems, too, are relatively young, far from saturated with species and extra susceptible to bio-invasions.

Policy is relevant as well. As a member of the European Union, the Netherlands has open borders with all other member states.

The Netherlands as a player in bio-globalisation

The Dutch have traditionally focused on trade, including overseas trade. In the 17th and 18th Centuries, the *Vereenigde Oost-Indische Compagnie* (Dutch East India Company, VOC), and the *West-Indische Compagnie* (West India Company, WIC) played an important part (see Boxes 72 and 73). The Netherlands for centuries has been a distribution country as well as a *bio*-distribution country, with beneficial as well as adverse consequences.

Beneficial effects included the introduction of various crops in Indonesia, such as the coffee plant[67], the cassava plant, the rubber tree and various vegetable crops from Europe. After the VOC had gone bankrupt in AD 1795, the Dutch also introduced the kina tree, from which the anti-malaria drug quinine is made.

In South Africa's Cape Province, the VOC managed to introduce citrus species, followed by the entire range of crops grown in Dutch vegetable gardens (see Box 13).

Adverse effects were:

- The introduction of pigs, cats, goats and dogs on islands in the Indian Ocean, which contributed to the extinction of many endemic species, including the famous dodo on Mauritius.
- Possibly introduction in America of yellow fever in Barbados in 1647 (see Box 73).
- Repeated imports of diseases into America that had already been introduced earlier and which were fatal to Native Americans.[68]
- Sailors spreading venereal diseases such as syphilis to every coast where goods were traded.

One disease is often assumed to have originated in the Netherlands, the Dutch elm disease. In fact, it was only discovered in the country, in 1917. The origin of the fungus is not known. The disease is caused by the fungus *Ophiostoma ulmi*, which is transmitted by the elm bark beetle. It has devastated elms in both Europe and North America. There is still no effective means of eradicating the fungus.

The Netherlands continues to be a hub for distribution, by road as well as water and air.

In addition, the following specific risk factors can be identified:

- The Netherlands is a substantial producer and major exporter of live material including livestock, vegetables, fruits, potted plants, cut flowers and flower bulbs. Included is living propagation material, mainly plant seeds, seed potatoes, sperm (especially of cows), broilers and piglets. The country even is the world's nr 1 exporter of broilers. Export of such propagation material represents a major risk because it ends up in agricultural systems all over the world. As regards ornamental plants and flowers: they are a well-known source of bio-invasions.
- The Netherlands exports large quantities of meat and dairy products, which also can transmit pathogens.
- The Netherlands has huge international flower auctions where living plant material from multiple parts of the world comes together, is traded, can come into contact, and is shipped out (Vermeulen *et al.* 2004).
- The Netherlands has many greenhouses for horticulture, and these have a warm microclimate. This means that bio-invasions can come not just from regions with a temperate climate, but also from tropical and subtropical areas. Within the Dutch climate there is little risk of spreading, but since Dutch horticulture is highly export-orient-

ed, the pest species involved can be further spread across the globe.

- Last but not least: the Netherlands is a major importer and (particularly) exporter of ballast water (Aquasense 1998a, see Box 76).

This does not mean, however, that the Netherlands carelessly spreads species around the world. If that were so, it would be quickly punished with trade bans. Therefore the country is highly active in preventing export of contaminated goods. Nevertheless, nearly every year one or more countries close their borders to Dutch flowers, plants, livestock, meat or dairy after infected goods are discovered. In recent years such bans were imposed on the Netherlands due to swine fever, bird flu, foot-and-mouth disease, Californian thrips and the Med fly. Like other trade-oriented nations, the Netherlands is, and will continue to be, a high-risk country where it comes to bio-invasions.

The Netherlands as an object of bio-globalisation

Considering the trade-oriented Dutch economy, it comes as no surprise that the country faces bio-invasions at an increasing rate.

Invasions of aquatic environments

Since about AD 1850, Dutch freshwater and salt-water bodies have seen a clear increase in the number of introductions (see Figure 8). Already, more than 200 exotic species have been reported, over half of which in marine and estuarine waters (see further Van der Velde et al. 2003, Bij de Vaate 2004 and Wolff 2005). An as yet unpublished species list by Van der Velde contains:

- in freshwater: 20 plants (including phytoplankton) and 65 animals (including protozoa);
- in marine and estuarine waters: 53 plants (including phytoplankton) and 131 animals (including protozoa).

For 36 of these species, it is not yet clear whether they will naturalise or remain limited to just one or a handful of specific locations.[69]

In recent years, particularly the past five years, the number of introductions has increased further, so the actual number of aquatic exotic species is now probably closer to 300.

It has been calculated that, from ballast water alone, one species will be introduced in the Netherlands every four years (Aquasense 1998b, referring to calculations in Aquasense 1998a). Only time will tell which of these species are there to stay.

The Rhine has seen a strong increase in the number of exotic invertebrate species over the last decades. These species first came mainly with ballast water from Eastern Europe and North America, and from 1992 particularly via the Rhine-Main-Danube canal (Van der Velde et al. 2003). Even more spectacular is the share of exotics in the number of individuals. Invaders – many of them Ponto-Caspians – now comprise more than 90%, and occasionally even more than 95% of the numbers of bottom-dwelling macro-invertebrate fauna. Densities of hundreds of thousands per m² were measured and swamping occurred. Although biomass was not systematically measured, we may safely assume that the invaders are dominant in that respect as well (Rajagopal et al. 1998, Tittizer et al. 2000, Haas 2001, Haas et al. 2003, Van Riel et al. 2006).

In short, a stealthy revolution has taken place in the invertebrate fauna within a few decades. Since this change is almost invisible to the layman and - unlike the *coup d' état* by the zebra mussel in North America - has done little harm so far, it has largely gone unnoticed. New species are still invading, and it may take decades for the fauna to become stabilised again.

Much better visible to the layman are the changes in marine and estuarine waters. Many

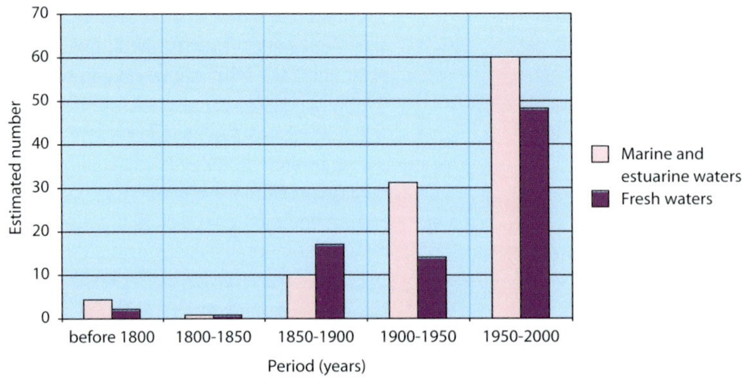

Figure 8.
Number of introduced species in Dutch waters over time.

Dutch beaches are today dominated by huge numbers of shells of the American jack knife clam *Ensis directus*, an invader that probably arrived in ballast water and was first observed in 1982 in the Netherlands. Likewise, at low tide enormous concentrations of Japanese oysters up to 0.5 m high can be observed in the Oosterschelde estuary and in parts of the Wadden Sea.

A key question for biodiversity is: will newcomers displace native species? So far there are few examples of this in the Netherlands; and sometimes these displaced species return – decades later, in some cases. One example is the American piddock *Petricola pholadiformis*, which appeared to be completely displacing the indigenous white piddock *Barnea candida*. Later this turned out not to be the case. However, the recent series of Ponto-Caspian invasions in the Rhine did cause some native species to disappear, possibly forever.

The mechanisms of such displacements are quite varied. One species will feed on a native species, another will compete it away and still another will occupy the substrate. Bij de Vaate (2004) lists a number of adverse effects of invasions by aquatic exotic species in the Netherlands. Complete displacement (extinction) hardly appears at all in this list, though. According to Wolff (2000) there are no known cases of extinction as a consequence of an introduced species in the marine environment, except for two species of marine birds that went extinct after the introduction of a terrestrial predator to their breeding habitat.

Of course, there are some known cases of extinction, and the number is increasing. In many cases, though, it is not clear if, and to what extent, the extinctions are to be attributed to bio-invasions. For instance, the indigenous crayfish *Astacus astacus* was decimated by the crayfish plague *Aphanomyces astaci*, a fungus imported with American crayfish. This provided room to other exotic species that had escaped from fish farms or had been released into the wild by aquarium owners. Of these species, the red swamp crayfish *Procambarus clarkii* is particularly resistant to crayfish plague. But the new species are also more resistant to water pollution. So the replacement of *Astacus* seems to be a combined effect of a bio-invasion and pollution.

Freshwater plants and animals

We now take a closer look at the composition of the group of invaders. As mentioned earlier, 20 exotic plants (including phytoplankton) and 65 exotic animals (including protozoa) have been found in freshwater. Figure 9 gives a breakdown by taxonomic group.[70] Obviously, flowering plants dominate among the exotic plants, while crustaceans and fish dominate among the exotic animals.

The geographical origin of these freshwater species is

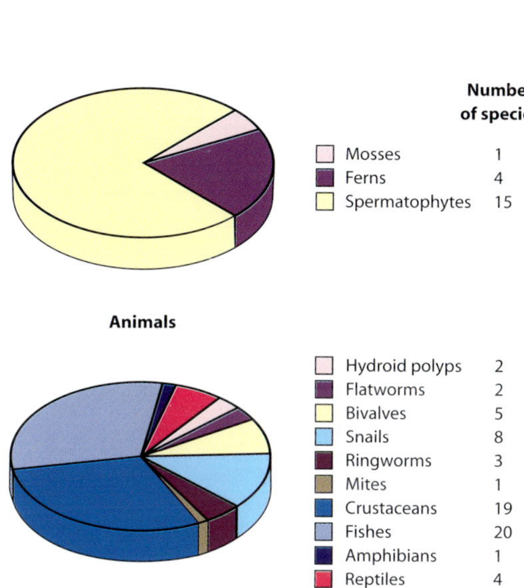

Figure 9.
Exotic *freshwater* plants and animals in the Netherlands, by taxonomic group.

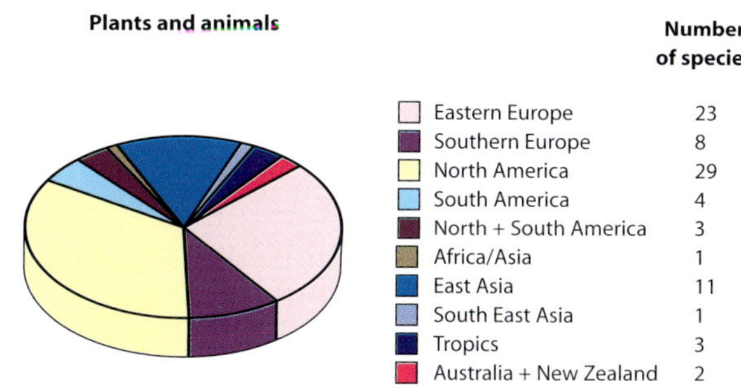

Figure 10.
Exotic *freshwater* plants and animals in the Netherlands, by geographical origin.

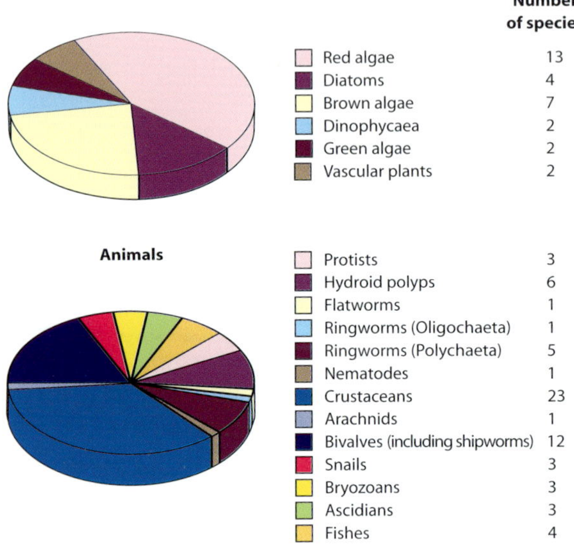

Figure 11.

Exotic *marine and estuarine* plants and animals in the Netherlands, by taxonomic group.

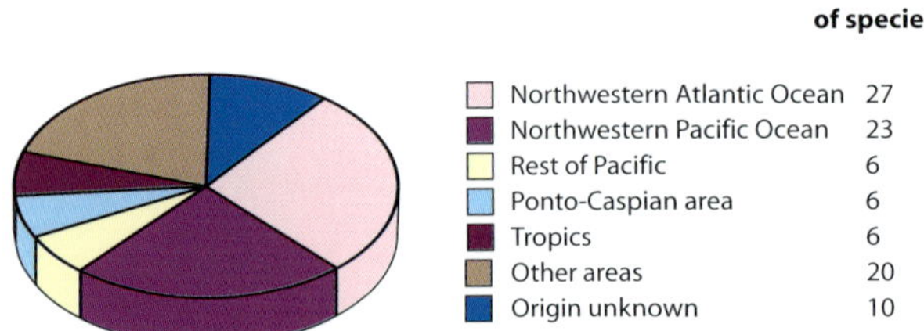

Figure 12.

Exotic *marine and estuarine* plants and animals in the Netherlands, by geographical origin.

given in Figure 10. Most species originate in Eastern Europe, North America and East Asia, reflecting climate zones on the one hand, and dominant trade routes on the other.

Some species are causing damage, such as the pumpkinseed sunfish *Lepomis gibbosus* (see Box 78).

Marine plants and animals

As stated, a total of 53 species of exotic plants (including phytoplankton) and 131 species of exotic animals (including protozoa) have been found in Dutch marine waters. Wolff (2005) lists 30 plant species and 68 animal species. The taxonomic distribution of these is given in Figure 8. The main difference between freshwater and marine plants is that red algae are the largest group among the latter plants; among marine animals, crustaceans and bivalves (rather than fish species) are the dominant groups.

Where did these species originate? This is outlined in Figure 9. Most species, by far, stem from the northwestern parts of the Atlantic and Pacific Oceans.

How permanent are these species settlements? Wolff considers 60 species to be permanent inhabitants of Dutch waters, while 18 are temporary and a further 18 are recent settlers. In addition, 4 species were imported but did not settle here; and for one species it is not known whether it has settled in the Netherlands or not. In addition to these figures, Wolff lists a further 12 species from the adjacent Northeastern Atlantic area which may either have been imported to the Netherlands or reached the Dutch shores on their own. He also lists another 38 species for which there is as yet little evidence that they are introductions. Finally, it should be noted that dozens more species have been found since this publication; their settlement status cannot yet be determined.

Plants

The wild flora in the Netherlands counts 1490 species of vascular plants (Odé *et al.* 2003). Of these, 1130 are indigenous species; a further 360 are not native but are naturalised. This latter group can be broken down as follows:

- 130 species settled in the Netherlands before AD 1500;
- 85 settled in the 19th century;[71]
- 145 settled in the 20th century.

Dutch botanists call species that settled in the Netherlands before AD 1500 (roughly the start of the Columbian Exchange) *archaeophytes*; those that settled later are called *neophytes*.[72] The rate of invasions is increasing. In the period 1500-1996 an average of one neophyte every three year became naturalised. However,

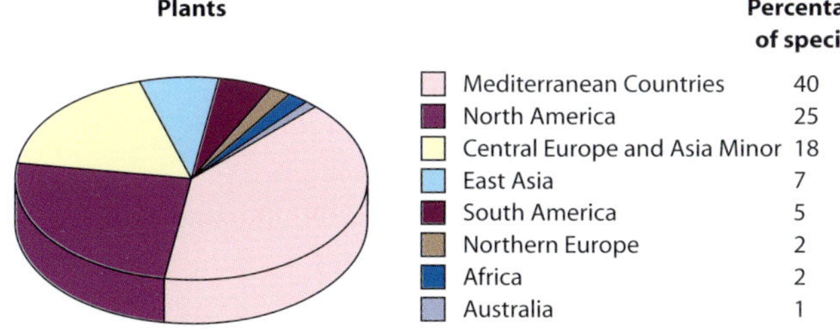

Figure 13.
Geographical and habitat origin of 125 plant species naturalised in the Netherlands since 1960 (Data from Denters 2004).

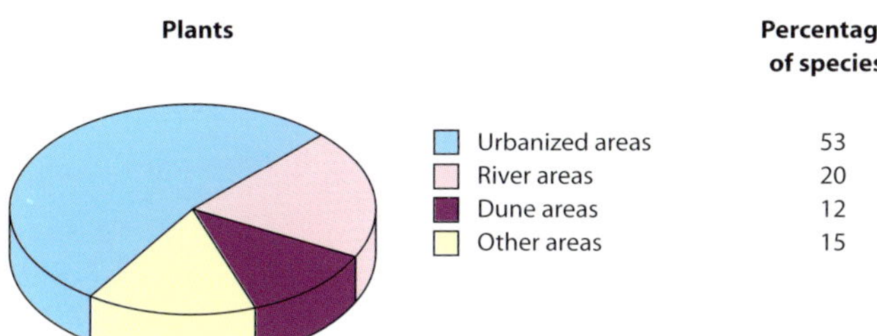

in the latest phase (1962-1996) it was two neophytes per year, which is six times faster (Pegtel 2003). These figures agree well with the findings of Tamis (2005).

Remarkably, no less than 65% of the newcomers have spread from gardens. Another 25% expanded their territory naturally from adjacent regions, and the final 10% were unintentionally introduced as stowaways with ships, trains, cars etc. The latter are known as "adventitious plants".[73]

Equally remarkable, more than half of the newcomers settled in urbanized environments, 60% of which on streets and 20% on brick walls (Denters 2004).[74] This makes the streets the foremost "reception area" of exotic species. That fits in to the pattern that invasions often take place in disturbed environments but it also has to do with climate change, which is most noticeable in stony environments. It is no coincidence that many newcomers stem from warm countries (see Figure 13).

The urban flora already consists of some 700 species – nearly half of the total Dutch flora! In the latest edition of the *Flora van Nederland* (Van der Meijden 2005), a new plant district has been added: urban areas. Of the 700 urban species, at least a third are exotics (Denters 2004).

Without a doubt, exotic species are contributing to the biodiversity in the Netherlands. In the urban plant guide by Denters (2004) as well as the tree guide for the south of Amsterdam by Lever (2002, see Box 74), exotic species are almost exclusively described as welcome acquisitions.

This is the sunny side of the story.

The shady side is that there is a significant chance that some of these newcomers will develop into pests. For example, some of the species in Amsterdam are spreading so fast that they have to be controlled, including the Caucasian wingnut *Pterocarya fraxinifolia* in the Beatrix Park (Lever, pers. comm.). The Japanese knotweed *Fallopia japonica* and the giant hogweed *Heracleum mantegazzianum* are swamping native plants locally. The latter species can also cause itching, blisters or even blood poisoning if touched by humans.

As for the damage in the country at large, Odé *et al.* (2003) have made up a list of the 47 most important invasive plants species. Of these, 21 cause ecological damage and 10 cause economic damage.

Ecological damage is mainly done by swamping and displacing native species in a range of habitats:

- Forest: most damage is done by four species from North America: black cherry *Prunus serotina*, locust tree *Robinia pseudoacacia*, red oak *Quercus rubra* and sycamore maple *Acer pseudoplatanus*. These species locally form dense monocultures and accumulate litter that is decomposed slowly, which harms other flora and perhaps fauna. Locally, the giant hogweed, too, is penetrating forests.
- Roadsides and wastelands: four species from Asia (including the Caucasus) are swamping on

Giant hogweed, native to the Caucasus, is swamping on an increasing number of locations in the Netherlands.

an increasing number of locations: Japanese knotweed, the giant knotweed *Fallopia sachalinensis*, the giant hogweed and the Himalayan balsam *Impatiens glandulifera*.[75]

- Dunes: heath star moss *Campylopus introflexus* swamps, replaces lichens, fixes the sand and thereby changes the habitat (Bal 2004).
- Heather: two of the species causing damage are native to North America: serviceberry *Amelanchier lamarckii* and high-bush blueberry *Vaccinium corymbosum*. The former invades dry heathland, changing the habitat; the latter invades moist heathland, replacing other herbs.
- Freshwater: the main invasive pests are four species from North America: floating pennywort *Hydrocotyle ranunculoides*, least duckweed *Lemna minuta*, water fern *Azolla filiculoides* and Nuttall's water weed *Elodea nuttalli* (though the latter species has shown a boom-and-bust pattern after its introduction). Where swamping, these species radically change the underwater light regime and replace native submerge species. Locally, Carolina flatwort *Cabomba carolinensis*, is causing similar problems.
- Salt-water: the introduced *Spartina anglica* (see Box 29) has almost completely displaced the indigenous *S. maritima* in Dutch mudflats.

Economic damage is mainly generated in agriculture, forestry and water management (Odé *et al.* 2003).

In agriculture, three important weeds were introduced in the early 20th century: disc mayweed *Matricaria discoidea* (native to North America and Northeast Asia), gallant soldier *Galinsoga parviflora* (Central America) and shaggy soldier *G. quadriradiata* (Central and South America). Chufa flat-sedge *Cyperus esculentus*, once a crop in ancient Egypt, was introduced by Arabs in Spain during the Middle Ages and from there spread across Europe. The species is a very noxious weed and is eradicated where possible. The expansion of maize since the 1950s was, and still is, followed by invasions of weed species of the genera *Amaranthus*, *Eragrostis*, *Panicum*, *Setaria* and *Sorghum*. Like maize, they are so-called C4-plants, which have a competitive advantage from the types of herbicides used in maize cultivation. This invasion process is still going on, also beyond maize fields.

As for forestry, the main invaders are four species from North America: black cherry, locust tree, red oak and sycamore maple. These species form dense monocultures and accumulate litter that is decomposed slowly, which harms other flora and perhaps the fauna.

As for water management, today the most damaging invader is floating pennywort (see Box 79)

Almost certainly more newcomers will cause damage in the future. Williamson's tens rule suggests that some 10% of all introduced species will become naturalised and of those 10% (i.e. 1% of all introduced species) will

Giant knotweed, native to Northeast Asia, is invading the Netherlands along with Japanese knotweed.

develop into a pest. However, as mentioned in Chapter 4, much higher percentages can be found in the literature as well. We recall that expansions may start only decades after the initial settlement. However, as for city plants, their opportunities to spread into the countryside will be limited by differences in microclimate: temperatures in the countryside are often several grades below those in cities. Outside the cities, at least two still rare exotics are suspected to cause trouble in the future: Australian swamp stonecrop *Crassula helmsii* in water courses and field dodder *Cuscuta campestris*, native to North America, in agriculture (Odé *et al.* 2003).

Land animals, birds and amphibians

The number of exotic vertebrates is much smaller than the number of exotic plants. From antiquity until the present, the total number of vertebrate invaders (not including fishes) was 12 species of mammals, 9 species of birds and 2 amphibians. An overview is given in Table 7. Based on this table, we can make some interesting observations:

- Of the total of 23 introduced species, 6 have spread widely in the Netherlands: the rabbit, brown rat[76] and domestic cat[77] as well as, more recently, the muskrat, coypu and Egyptian goose[78].
- 18 Introductions date from the 20th century, only five happened previous to the 20th century. That, however, does not necessarily mean that the rate of introductions has increased rapidly, because there may have been older introduc-

tions that ultimately failed and are not included in the list.

- Most species originate in temperate climate zones in the Northern Hemisphere. Only two species (the coypu and black swan) are from the Southern Hemisphere. Four (coypu, rose-ringed parakeet, Egyptian goose and house crow) are tropical.
- 16 species are from the Old World, of which 3 or 4 (rabbit, mouflon, Italian crested newt and possibly fallow deer) are from Southern Europe; 7 species are native to the New World.[79]
- No less than 20 of the 23 species were intentionally introduced, particularly for production, hunting or ornamental purposes. That is comparable to the situation in Switzerland, where 35 of 37 vertebrate animals were deliberately introduced (Wittenberg 2005).
- Nearly all birds were introduced for ornamental reasons (particularly for water bird collections), while most mammals were introduced for fur production and/or hunting.
- Three of the mammals were deliberately introduced elsewhere in Europe, then reached the Netherlands on their own: the muskrat, raccoon and raccoon dog.
- Only two species (the brown rat and, more recently, the house crow) hitched a ride with ships, illustrating our previous statement that large animals rarely hitch a ride but are deliberately introduced.

Introduced species of land animals, birds and amphibians in the Netherlands*

Taxon	Hitched a ride on or was imported for	Naturalised since	Origin	Damage/Risk
Mammals (12 species)				
Domestic cat	pet	antiquity	N Africa/S Asia	predation on wild fauna
Fallow deer	hunt, production	antiquity	SE Europe/SW Asia	overgrazing
Rabbit	production, hunt	13th century	SW Europe	feeds on crops, digs holes
Brown rat	hitched a ride on ships	17th century	E Asia	vector for pathogens, feeds on crops, gnaws at cables etc.
Mouflon	hunt	1909	Corsica/ Sardinia	competition?
American mink	production	1929	N America	feeds on crops
Coypu	production	1935	S America	damages riverbanks
Muskrat**	production	1946	N America	digs holes in dams and dikes
Raccoon**	production	1960	N America	competition
Siberian chipmunk	zoos	1972	E Asia	
Raccoon dog**	production, hunt	1986	E Asia	competition?
Reeves' muntjac	ornamental	1998	China/Taiwan	feeds on forest vegetation, competition
Birds (9 species)				
Pheasant	ornamental, hunt	500-800	W Asia	
Mandarin duck	ornamental	1964	E Asia	
Ruddy duck	ornamental	1973	N America	
Egyptian goose	ornamental	1967	Africa	competition
Canada goose	ornamental	1977	N America	feeds on crops, flattens river banks, causes eutrophication
Bar-headed goose	ornamental	1977	NE Asia	
Black swan	ornamental	1985	Australia	
Rose-ringed parakeet	ornamental	ca. 1968	Africa/S Asia	competition
House crow	hitched a ride on ships	1997	S Asia	infections, competition?
Reptiles (0 species)				
Amphibians (2 species)				
Bullfrog	gardening centres, ponds	1980-90	America	competition
Italian crested newt	gardening centres, ponds	1997	Italy	competition
Total: 23 species				

*We applied a somewhat different definition than did Bal (2004). We also included introductions from before 1850 but of the birds only those species breeding in the wild.
** First introduced elsewhere in Europe, then released or escaped, and finally spread spontaneously to the Netherlands.

Sources: Bal (2004) supplemented from Lange et al. (1994), Nowak (1999), Van Dam (2000) and SOVON (2002)

- Four species, all mammals, are causing large-scale damage: the domestic cat[80], rabbit, brown rat and muskrat.[81] The Egyptian goose also seems to be developing into a pest, threatening some indigenous bird species.[82]

The picture sketched above may be a bit too bright since Bal (2004) has insufficiently taken into account the possible effects of parasites of the invasive species. In the UK, for instance, a parasite of the pheasant has proved harmful to the partridge; the raccoon dog is spreading the fox tapeworm *Echinococcus multilocularis*, dangerous to humans; the domestic cat (not considered exotic by Bal) transmits toxoplasmosis to cattle and to humans; and the fallow deer (as well as the indigenous roe deer) can transmit ticks carrying the *Borrelia* bacterium that causes Lyme disease (see Box 68).[83]

Arthropods

As for the arthropods, an inventory was recently made of invasive species in the Netherlands (Reemer 2003). This included groups such as flies, butterflies, beetles, dragonflies, spiders, mites and true bugs. The inventory is limited to invasions from 1950 until 2005, and is also incomplete in other ways: taxonomic groups for which no experts could be found were omitted; also excluded were: strictly aquatic species; species that have not expanded very much; and older invasions, such as those by various kinds of cockroaches.

On the other hand, European species that actively[84] settled in the Netherlands, i.e. without any human assistance, were included in the inventory – unlike the previous overview, which omitted such species.[85] The figures therefore are not really comparable.

The total number of species is 104. Disregarding active settlers, as we largely do in this book, the total amounts to 67. So, in the past half of a century, the number of invasive arthropods has been three times as high as the total number of invasive vertebrates (excluding fish species) in *all* of history.

A closer analysis of the 104 species reveals some interesting patterns:

- 73% species are from Europe, 27% are from elsewhere.
- The non-European species are from all parts of the world except Antarctica; relatively many (11%) are from North America.
- The European species have largely (88%) settled actively, i.e. without any human assistance.
- The non-European species have all settled pas-

The house crow, native to Southern Asia, reached the Netherlands by hitchhiking on a sea-ship.

sively, though some of them initially settled elsewhere in Europe and then came to the Netherlands on their own.

These developments are happening at an alarmingly fast rate. Since 1992 alone, already 30 species have settled and spread widely; that is an average of 3 species per year. We have examined this group of 30 species in more detail:

- 19 species are from Europe, 7 directly or indirectly from North America and 1 each from South America, Asia, Africa and Australia.
- The majority of the species is from the Northern Hemisphere.
- Most settlements in Europe are active and probably caused by climate warming.
- The most important vector involved in passive distribution is the import of plant material (7 species). Other vectors are: the import of apples, tobacco and dogs; introduction as biological pest control; and shipping (all 1 species each).
- Only a single case of passive distribution was an intentional introduction (namely for biological pest control); all other cases were accidental introductions of "stowaways." That is the mirror image of what was found in vertebrates and fits the pattern that most small invasive species "hitch a ride" while most large species are deliberately imported.
- Of the 11 non-European species, one (the andromeda lace bug *Stephanitis takeyai* from Japan) was - probably not by chance - first found near the town of Boskoop, a key player in the global tree cultivation industry.

Unfortunately, the inventory does not state how these figures relate to the total number of species in these groups in the Netherlands, i.e., what *proportion* of the species is exotic.

Also interesting are the data on "usefulness" and "harmfulness":

- Of all 104 invasive species, 7% are considered useful and 36% harmful.
- Of the European species, only 22% are harmful as against 75% of the non-European species.
- Of those 30 invasive exotics that have settled since 1992, as many as 23 (77%) are considered harmful or potentially harmful to society and/or nature.
- Damage is done or expected to agriculture, forestry, ornamental plants and trees; human habitation and places of work; and human and animal health. Most damage is from feeding on plants. Other damaging effects are skin irritation in livestock and pets; transmission of diseases;

and even damage to old books and zoological collections.

- Little is known about the effects on nature. There is not yet a case of an invasive arthropod strongly affecting an ecosystem. However, one grasshopper species and one species of harvestman are suspected to be displacing native competitors.

In 2003 the Western corn rootworm *Diabrotica virgifera* was first reported near Amsterdam Airport. This beetle is native to Mexico and Central America. In the US it has ominously been named the "billion dollar bug". In 1992, it was first found in Europe, more specifically in Belgrade, from where it spread. It is now threatening the cultivation of maize across Europe because its larvae feed on the roots of the corn plant. In the Netherlands, the beetle has been found ever more often since 2003, not just near Amsterdam Airport but also in the vicinity of Maastricht/Aachen Airport, indicating multiple independent settlements in the Netherlands. It appears that an invasion of this species is only a question of time.[86]

There have been recent reports that indigenous ladybug species of the genus *Coccinella* are being displaced by the Asian ladybug species *Harmonia axyridis*, which had been introduced in Belgium and France to combat lice in greenhouse horticulture.[87]

An even more recent find is the Japanese beetle *Phloeosinus rudis*, which affects ornamental conifers. The beetle was spotted in Ridderkerk in 2004.

When examining the figures mentioned, it should be noted that these numbers include only species that have spread *widely* in the Netherlands.

The fact that exotic species from distant countries are often more damaging than species originating in more nearby areas can be explained with the "enemy release" hypothesis (see Box 28): species that come to the Netherlands as a result of climate change will be joined by their natural enemies more often than species from distant regions. But this does not apply to viruses, which have few natural enemies. That helps explain why there are some real killers among the viruses that are advancing on the Netherlands as a result of climate change, such as the bluetongue and West Nile viruses (see Box 86).

Nematodes

Nematodes are often overlooked in surveys of invasive pests. Yet they are an important cause of crop losses as well as pesticide use. Key species in the Netherlands are the potato cyst nematodes *Globodera pallida* and *G. rostochiensis*, introduced in the mid-1880s (see Box 45). Several other species have been introduced as well. They are difficult to control, especially in narrow crop rotations.

Harmonia axyridis, a ladybug form Asia introduced in Belgium and France for lice control, is invading the Netherlands while displacing indigenous ladybugs.

Economic costs from bio-invasions

How much economic damage do these exotic species cause? We will examine this sector by sector.

Fishery and aquaculture

Bij de Vaate (2004) lists 36 aquatic exotic species with clearly negative effects. Most of those also have economic effects, particularly in:

- Fishery: shellfish poisoning, parasites, damage to fishing nets.
- Aquaculture: competition for food and territory, parasites.
- Water management: clogged waterways.
- Industry: clogged cooling water intakes.

As yet, however, few economic figures are available to specify this.

Agriculture and forestry

Dutch agriculture and horticulture harbour numerous harmful invasive species. This involves diverse taxonomic groups and crops (see Table 8).

In addition, insects and mites cause damage to ornamental plants, in gardens and parks as well as ornamental plant cultivation, including chestnut, lime tree, maple, ornamental conifers, hollyhock, hydrangea and heather.

The total costs of chemical pest control in Dutch agriculture and horticulture amount to some €300 million per year since 1990 (LEI 2003b). Added to that must be the costs of pest control equipment, biological pest control in greenhouses, monitoring, weather monitoring stations and product sampling. An arbitrary estimate of the costs is €50-100 million per year, upping the total costs to some €350-400 million a year. This does not include phytosanitary measures on the border.

Neither the *Plantenziektekundige Dienst* (Plant

Table 8.
Some harmful invasive species in Dutch agriculture and horticulture

Taxonomic group	Example	Crop
viruses	potato virus Y	potato
	tomato spotted wiltvirus	peppers/tomato
bacteria	ring rot and brown rot	potato
fungi	*Phytophthora infestans*	potato
nematodes	*Globodera pallida* and *G. rostochiensis*	potato
insects	beet armyworm	greenhouse horticulture
mites	carmine spider mite	tomato, carnation and
		other ornamental crops
plants	chufa (tiger nut, yellow nutsedge)	corn, flower bulbs
	black cherry	forestry

Protection Service) nor the *Landbouw Economisch Instituut* (Agricultural Economics Research Institute) has information available as to the extent to which invasive species contribute to the costs of pest control. One of the few sources we were able to find is a publication by Van Lenteren *et al.* from 1987, with figures on invasive insects in horticulture (see Table 9). The authors noted that invasive insects do not cause any problems in the cultivation of vegetables in soil, but all the more in orchards and greenhouses. They also observed considerable problems in forestry, city parks and the storage of food in silos and houses.

We estimate (again arbitrarily) that 40-60% of the costs of pest control can be attributed to invasive species (using the year 1800 as cut-off point, so *Phytophthora*, for instance, is considered exotic). Thus, the total costs of pest and weed control in agriculture and horticulture are an estimated €156-240 million per year. In forestry, the costs of combating the black cherry are €2 million per year[88]

Additionally, there are substantial economic *losses and damage* that invasive species can cause despite prevention and control efforts. The damage can be quantitative and/or qualitative. Hardly any figures for this are available.

One of the few sources is the *Fauna Fonds* (Fauna Fund), which compensates farmers and others for damage done by a small group of organisms: birds and mammals including wild geese and wild boar. In 2003, native species accounted for the lion's share of the €6 million for compensation paid to farmers (and others). Exotic

species like rabbit, fallow deer, Canada goose and pheasant together accounted for a mere €17,000.

These figures are not representative, however, since the damage involved is very low compared to the damage done by groups such as weeds, insects, fungi and microbes. The same may hold for the share of exotics. In addition, some viruses and parasites may be spread by wild birds and wild boar, but such risks are not compensated for by the Fauna Fund and do not figure in their data.

Judging from figures from the US (Pimentel *et al.* 2005) and Australia (Canyon *et al.* 2002) the losses and damage are 3 to 14 times as high as the control costs. We prefer to err on the side of modesty and estimate the damage at between 100% and 300% of the control costs.[89] That equals €117-720 million, raising the total costs of pest control plus losses and damage to €234-960 million per year.

In addition, out of a total of €28 million in institutional costs by the *Plantenziektekundige Dienst* (PD, Plant Protection Service), 70% to 80% goes to preventing bioinvasions by quarantining organisms (L. Smits, personal communication). That equals €21-24 million per year.

To this can be added part of the equipment costs of the *Algemene Inspectiedienst* (AID, General Inspection Service) as well as the *Rijksdienst voor de Keuring van Vee en Vlees* (RVV, Inspection Service for Livestock and Meat). A small fraction of the research and extension budget of the Ministry of Agriculture, too, can be attributed to the prevention and control of bio-invasions.

Some invasive insects in Dutch horticulture and pest control costs in the 1980s

Species	Affected crop	Control costs (x million Dfl/year)	Method
Trialeurodes vaporariorum	tomato	4.5	biological
Chrysodeixis chalcytes	vegetables	2	biological
Liriomyza trifolii	chrysanthemum	3.5	chemical
	gerberas	1.7	chemical
Adoxophyes orana	apples and pears	1.8	chemical, non-selective
Adoxophyes orana	apples and pears	3.3	chemical, selective
Eriosoma lanigerum	apples	1	chemical

Source: Van Lenteren et al. (1987)

Livestock husbandry

Over the last decade the Netherlands was struck by three large epizootics caused by invasive animal viruses:

- Classical swine fever in 1997.
- Foot-and-mouth disease in 2001.
- Bird flu (caused by AI virus subtype H7N7) in 2003.

In addition, there was a stealthy invasion of a new emerging disease: Bovine Spongiform Encephalopathy, better known as BSE or "mad cow disease". This is not caused by a virus but by a protein, a so-called prion. Since BSE is a zoonosis that can cause a deadly disease in humans, panic broke out for some time.

The swine fever virus was introduced from Germany, the foot-and-mouth virus from Asia (via Britain and France) and the BSE prion probably from Britain (via feed or cattle imports). Where the bird flu virus originated is still unclear (see Boxes 67, 80). We have assumed it was introduced.

The preferred strategy in the EU to control outbreaks of highly contagious livestock diseases is rapid containment and eradication. This is to prevent spreading of the disease to other member states, and to prevent the pathogen to become endemic. Containment is pursued by immediately imposing a standstill for transport of livestock as well as for export of animal products from the infected region. Eradication is pursued by culling and destroying infected (or possibly infected) animals. Vaccines could make a more humane alternative, but even if these are available, they are often not used for economic reasons. Producers fear that vaccinated animals might still carry and spread the virus (which often is not the case) and that trading partners might close their borders for vaccinated livestock and their products. They also fear that supermarkets might ban the products, assuming that consumers will not fully trust them. Hence during an outbreak many more animals are culled than is biologically required to control the pathogen.

BSE was a somewhat different story. Since the prion is not infectious, the number of animals culled was limited. On the other hand, long-term costs were generated by a ban on the use of so-called risk organs, a ban on the use of bone meal in feed, and standard testing of animals at the slaughterhouse

Considering all this, it does not come as a surprise that all four outbreaks generated huge costs: direct costs of culling and destroying animals; and indirect costs from export bans and reduced domestic consumption. Such bans hit the Dutch economy relatively hard, since the livestock industry is highly export-oriented.

The costs have been calculated profoundly. Appendix 3 gives an overview. It is presented in this book as an example of how to calculate the incredibly complex economic impact of an invasive livestock disease which is either highly contagious or a zoonosis. The calculated costs were (see also Table 10):

- €1.5-1.9 billion from the swine fever virus;
- €874 million from the foot-and-mouth virus;
- €880 to 980 million from the bird flu virus;

- €554 to 940 million from the BSE prion in the period 1995 to 2005.

The costs from the most expensive outbreak, swine fever in 1997, amounted to 0.5% of the Dutch GDP (see Appendix 3).

Adding up the costs from all four epizootics in the period 1995 to 2005, we arrive at the huge sum of €3.9 to 4.6 billion.

It would be too simple to translate this simply into €390-460 million annually, since the circumstances of the decade may have been a tragic coincidence. Hence we have used 460 million per year as an upper limit. This includes €21 million damage done to other industries, mainly tourism. As a lower limit, we have, somewhat arbitrarily, used 25% of €390, approximately €100 million per year (see Appendix 3).[90]

The permanent threat of epizootics generates institutional costs in every country. In the Netherlands, the most important institutions are the *Algemene Inspectiedienst* (AID, General Inspection Service), the *Rijksdienst voor de Keuring van Vee en Vlees* (RVV, National Inspection Service for Livestock and Meat), and the private *Gezondheidsdienst voor Dieren* (GD, Animal Health Service). We have, again somewhat arbitrarily, attributed 10-30%, 20-60% and 20-40%, respectively, of the annual capacity costs of these institutions to the prevention and control of invasions of livestock and plant pathogens combined.

Nature and landscapes

Regarding the effects on nature conservation, we for brevity's sake refer to the species and effects mentioned earlier.

In the Netherlands only a handful of those exotic species that are harmful to nature are actively combated (disregarding the muskrat and coypu, which are primarily harmful in other aspects). We mentioned the black cherry, Reeves' muntjac and the fallow deer. Of these, only the black cherry has thus far claimed high control costs.

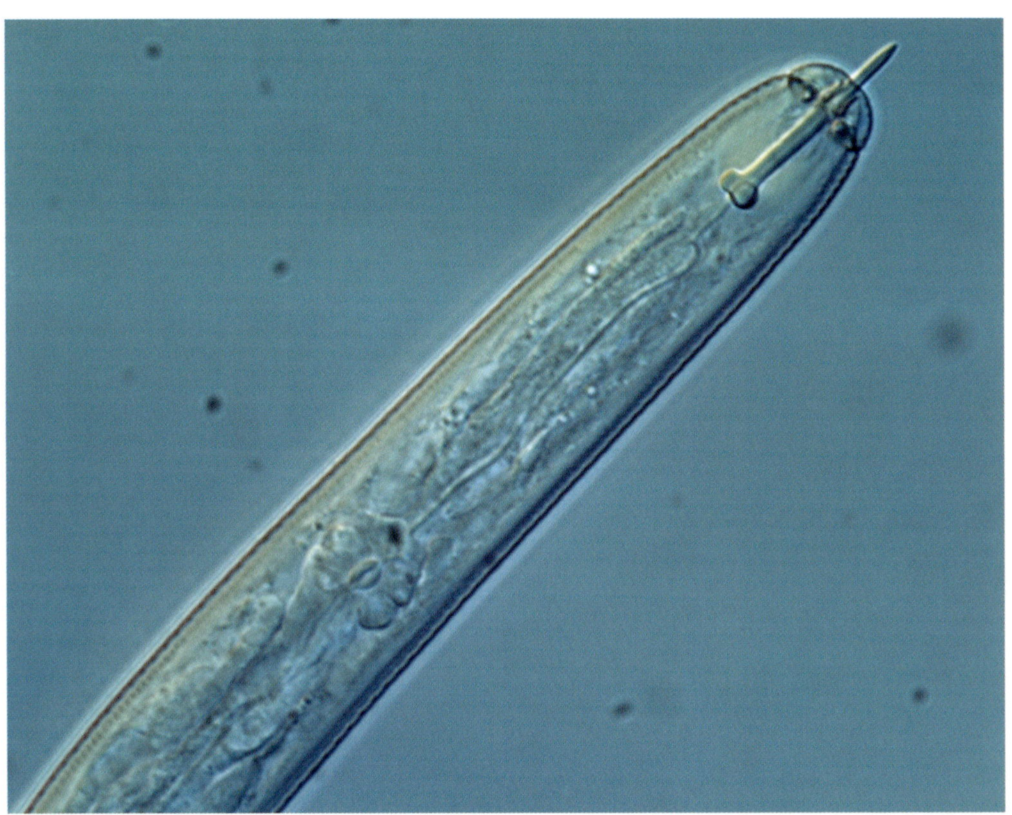

The potato cyst nematode, native to the Andes region, was first found in the Netherlands in the mid 1880s. It still is a major target for pesticide use.

A newcomer with potentially major consequences is the rabbit hemorrhagic disease virus, a calici-virus causing fatal haemorrhaging in rabbits. The virus was first reported in China in 1984 and in the Benelux in 1990. Recently, the number of rabbits in the Netherlands has dropped dramatically and the rabbit has even disappeared from some areas altogether (followed by a recovery in some areas). Locally this causes vegetation to grow high, to the displeasure of nature managers. However, an alternative is available. Already before the appearance of the rabbit virus, grazing animals such as horse and cow races were introduced in many Dutch nature reserves to keep the vegetation open and varied (although there has been some discussion whether this really benefits biodiversity). Thus it seems that the ecological damage from the virus will be limited largely to those reserves where large grazing animals are absent.

An upcoming problem is the rapid proliferation of exotic goose species, mainly Egyptian goose, Canada goose *Branta canadensis* and bar-headed goose *Anser indicus*.[91] It is too early to say what the ecological consequences will be, though the Egyptian goose is aggressive to other birds. However, the damage from these exotics is overshadowed by the explosive increase of the indigenous greylag goose *Anser anser* since 1961, which has become a real pest to cattle and sheepfarmers.

Recreation and tourism

Without seeking completeness we mention four effects from exotics on recreation and tourism:

- Incidental invasions of toxic algae can hurt tourism, causing an estimated €1 million in damage to the tourist industry.
- The Japanese oyster impedes beachside recreation at some locations along the Oosterschelde (East Scheldt) estuary.
- The rabbit hemorrhagic disease virus (mentioned above) has drastically reduced opportunities for hunting rabbits. The economic damage of this, in terms of hunting rights, has as far as we know not yet been quantified.
- Many roads and paths were closed off for several months during the foot-and-mouth disease outbreak of 2001. The income loss of cafés, restaurants and hotels accounted for as much as €200 million.

Townscapes

The urban vegetation is enriched as well as threatened by bio-invasions. As mentioned, the flora is spectacularly diversified by plant introductions. So far, only few among the newcomers have caused damage. Almost certainly there are associated control costs but these are not quantified.

The Egyptian goose invaded the Netherlands after it escaped and was released from enclosed bird ponds.

"Dutch" elm disease causes breaches in the rows of elms bordering the canals of Amsterdam. The causative fungus (above) is *Ophiostoma ulmi*.

On the other hand invasive plant pathogens have done much damage to urban trees and townscapes. The fungus *Ophiostoma ulmi*, transmitted by beetles, causes Dutch elm disease and has killed thousands of elms *Ulmus* spp. in parks and streets, including many along the famous canals of Amsterdam. Damage and control costs are considerable but not quantified.

Recently another characteristic city tree has come under attack from an invader. The horse chestnut *Aesculus hippocastanum* – imported from Turkey in AD 1550 – is since 1998 being attacked by the horse-chestnut leaf miner *Cameraria ohridella*. Trees start colouring brown in July but do not die. The parasite was first found in Macedonia in 1985, but its native region is not yet known. That makes it more difficult to find an appropriate natural enemy. So far, the damage and control costs are small but they will increase if trees become so damaged that they have to be cut and replaced.

Houses and other buildings
Regarding houses and other buildings, the most important invaders are some exotic species of cockroach as well as the house longhorn beetle *Hylotrupes bajulus*, the pharaoh ant *Momorium pharaonis*, the house cricket *Acheta domestica*, the skin beetle *Dermestes haemor-* *rhoidalis* and the brown dog tick *Rhipicephalus sanguineus* (see Box 82).

Cockroaches are a main cause of asthma, with an estimated treatment cost of €30-60 million per year. We assume that exotic cokckroaches are responsible for 50% of these costs. We have no figures for the costs of the other species mentioned.

Environment
Pesticides and herbicides are used to prevent bio-invasions and to control invaded species. For instance, chlorine is used to control plants clogging cooling water intakes. Currently the herbicide MCPA is used in some waterways for combating floating pennywort.

In agriculture, much of the pesticide and herbicide used can be attributed to exotic species like the fungus (or rather oomycote) *Phytophthora infestans* and the potato cyst nematodes *Globodera pallida* and *G. rostochiensis*. Even soil disinfection is used to control these nematodes. Environmentally critical substances such as dichlorvos (DDVP) are being used to eradicate quarantined invaders such as *Thrips palmi*.

The Netherlands is a major exporter of living plant material. In order to comply with the strict phytosanitary standards of many importing countries, growers and

Horse chestnuts start colouring brown as early as July after infection by the horse-chestnut leaf miner, an insect of unknown origin.

exporters apply much pesticides, especially on non-food products (like flowers and flower bulbs) and propagation material (seeds and seed potatoes) where residue tolerances are high. In some cases this even includes methyl bromide, a chemical harming the ozone layer. More often this chemical is applied on wooden pallets and packaging woods (Vermeulen & Kool 2006).

The environmental effects are difficult or impossible to express in monetary terms.

Public health

Every year many pathogens are imported into the Netherlands, but only a few are able to survive for a longer period of time. The most important of these are HIV/AIDS and influenza. Other imported diseases such as malaria, typhoid fever, Ebola fever and Lassa fever are dangerous, but never more than temporary visitors in the Netherlands. Bird flu occasionally harms humans. During the 2003 epizootic caused by the H7N7 subtype, 89 humans were infected, one of which died. However, the virus was soon eradicated in a campaign combining stringent hygiene measures and mass culling of poultry.

In recent years the HIV virus caused less than 50 human fatalities per year (25 in 2005). The virus generates substantial treatment costs as well as productivity losses. Influenza viruses cause some hundreds of fatal cases per year, which can rise to several thousands in some years (see Appendix 4). In addition, these viruses cause very high productivity losses due to the large number of people contracting the disease.

The estimated annual costs from the **HIV virus** are **€103–310 million** (see Box 84).

The estimated annual costs from **influenza viruses** are **€429–841 million**, almost entirely due to productivity losses (see Box 83).

The costs from the flu would go sky high if the present H5N1 virus mutates into a highly virulent and contagious lineage and causes a **severe pandemic**. In that case, costs may amount to **€22 billion. Assuming a GDP of €505,6 billion per year (2005), that would make 4.4% of GDP**, well above the 3% estimated by the World Bank for the global economy. The open Dutch economy will be hit relatively hard by the collapse of international traffic and trade that is expected to follow a pandemic. The economy will shrink, and recovery may be slow.

Compared to the huge impacts from these two viruses, the health impact of other invasive species is very small.[92] For the sake of completeness, we mention a few:

- Cockroaches are an important cause of asthma. The costs of treatment alone are an estimated €30-60 million per year (see Box 82).

Overview of the economic costs from invasive species in the Netherlands

Costs in millions of euros
*In **bold**: annual costs. Not bold: incidental costs.*
In parentheses: losses and damage estimates based on prevention + control costs multiplied by 1 (low estimate) and 3 (high estimate).
NA = Not Available

Sector, invader and type of damage	Costs of prevention & control	Losses & damage	Total costs
Agriculture & horticulture			
Pathogens, pests and weeds (40 - 60% exotic)	**117–240**	**(117–720)**	**(234–960)**
Forestry			
Black cherry	**2**	**(2–6)**	**(4–6)**
Animal husbandry			
BSE (1997-2005)	519 – 769	35 – 171	554 - 940
Foot-and-mouth disease (2001)	432	233	665
Classical swine fever (1997)	1,400	135 – 462	1,535 – 1,862
Bird flu (2003)	770	100 – 200	870 - 970
Multi-year average for all invasive diseases	**78–337**	**12.6–106.4**	**90.6–443.4**
Bee-keeping			
Varroa mite	NA	NA	NA
Fishery & aquaculture			
Japanese oyster	NA	> 1	> 1
Toxic algae (fishery & shellfish industry)	**4**	**(4–12)**	**(8–16)**
Chinese mitten crab damaging fishnets	–	NA	NA
Shipping & boating			
Japweed (wireweed)	–	NA	NA
Water management			
Muskrat	**31**	**(31–93)**	**(62–124)**
Coypu	**0.8**	**(0.8–2.4)**	**(1.6–3.2)**
Shipworm damaging poles and sheetpiling	NA	NA	NA
Floating pennywort	**2–4**	**(2–12)**	**(4–16)**
Energy sector & industry			
Water plants	NA	NA	NA
Tubeworm *Ficopomatus enigmaticus* clogging cooling water intakes	NA	NA	NA
Zebra mussel clogging cooling water intakes	NA	NA	NA
Nature conservation			
Fallow deer	NA	NA	NA
Reeves' muntjac	NA	NA	NA
Foot-and-mouth disease causing animal culling	1	NA	1
Urban environment			
Dutch elm disease	NA	NA	NA
Horse-chestnut leaf miner	–	NA	NA
Private gardens			
Various species	NA	NA	NA

Sector, invader and type of damage	Costs of prevention & control	Losses & damage	Total costs
Houses and other buildings			
Pharao ant	NA	NA	NA
Argentine ant	NA	NA	NA
Cockroaches	NA	NA	NA
House longhorn beetle	NA	NA	NA
Public health			
HIV virus	**50–150**	**53–160**	**103–310**
Influenza viruses	**38–59**	**391–782**	**429–841**
Cockroaches (50% exotic) causing asthma	NA	**15–30**	**15–30**
Clinging jellyfish (doctor's diagnosis and medication)	NA	NA	NA
Bird flu causing conjunctivitis and lung problems	NA	NA	NA
Fallow deer reservoir for Borrelia (Lyme disease)	NA	NA	NA
Recreation & tourism			
Algae from ballast water causing nuisance and harm	NA	NA	NA
Toxic algae harming tourism	-	NA	NA
Foot-and-mouth disease harming hotels & catering	-	**5–21***	**5–21***
Rabbit haemorrhagic disease harming rabbit hunting	NA	NA	NA
Research and surveillance**			
Ministry of Agriculture, Nature & Food Quality,			
5% - 10% of research and education budget	**42–83**	-	**42–83**
Faculty of Veterinary Medicine,			
Utrecht University, 10% - 20% of budget	NA	-	NA
Plant Protection Service,			
70% - 80% of institutional costs	**21–24**	-	**21–24**
Inspection Service for Livestock and Meat,			
20% - 60% of institutional costs	**21–74**	-	**21–74**
National Animal Health Service,			
20% - 40% of institutional costs	**9–18**	-	**9–18**
General Inspection Service,	**5–16**	-	**5–16**
10% - 30% of institutional costs			
Other relevant institutional costs	NA	-	NA
Total annual costs (not including NA items)	**461–1044**	**673–1945**	**1134–2989**
Estimated total annual costs of all NA items			**20–50**
Total annual costs (including NA items)			**1154–3039**

** A small fraction of these costs are to be attributed to the building and construction industry and to the transport business.*
*** The total equipment costs are based on figures from the 2005 Dutch State budget; figures are for 2004 but for those of the General Inspection Service (AID), which are for 2005. Percentages are based on rough estimates and, in the case of the Plant Protection Service (PD), on a personal communication by L. Smits.*

- The brown dog tick *Rhipicephalus sanguineus*, originally from Southern Europe, can transmit parasitic diseases like babesiosis and filariasis, occasionally even to humans (see Box 82).
- The giant hogweed *Heracleum mantegazzianum* causes itching, blisters and blood poisoning.

Total costs

There are no figures available for the total amount of damage caused by invasive species in the Netherlands. We have made a first, crude attempt, in anticipation of more thorough studies. The goal is merely to estimate the magnitude. There are at least two methods to arrive at an estimate: 1) extrapolating foreign data, and 2) adding up the individual cost items.

Method 1: Extrapolation

Extrapolation can take place in three different ways:
- Based on the *surface area* of the Netherlands, compared to the surface area of countries for which figures are available, assuming that the average amount of damage by invasive exotic species per square km is the same in every country. This does not appear to be a fruitful approach since the Netherlands are densely populated with people as well as livestock; invasions can thus cause more damage per square km than in other countries. In addition, yields per hectare in agriculture are relatively high because the Netherlands has favourable soils and climate.
- Based on the human *population size* in the Netherlands, compared to the population size of other countries, assuming that the damage is proportional to the size of the population. However, there is little to support this assumption. For one, much of the damage occurs in agriculture, forestry and animal husbandry, and the size of these sectors, per inhabitant, differs significantly among countries.
- Based on the *gross domestic product* (GDP) of the Netherlands, compared to the GDP of other countries, assuming that the damage is proportional to GDP. This assumption, however, is not very plausible either.

Furthermore, none of these three methods takes into account the fact that the Netherlands is a distribution country, into which many exotic species are imported. In addition, as we observed, the total sums given by Pimentel *et al.* (2002) are largely underestimates, since data for many items are missing.

If we momentarily ignore these concerns and use the 5% of GDP given by Pimentel *et al.* (2002), we can calculate as follows. According to the *Centraal Bureau voor de Statistiek* (CBS, National Statistics Agency), the GDP of the Netherlands was €505.6 billion in 2005. Taking 5% of that, we get an unbelievably high figure for costs from invasive species: €25.3 billion per year.

If, on the other hand, we use population size, our total amounts to 16,3 million people x $240 = $3.84 billion per year. At a conversion rate of $1 to €0.78, that would equal €3.0 billion per year. This figure seems more realistic, although it is not better founded.

Method 2: Addition and estimation

The second method has its problems too. There are more gaps than actual data (see the numerous *NA = not available* items in Table 10). Some major missing items are the production losses from invasive pathogens and weeds in agriculture and horticulture. As mentioned earlier, we have estimated these losses at 100% to 300% of control costs. The same method was used to estimate the damage caused by the muskrat and coypu. Thus we arrive at the following sums:
- the total costs of preventing and controlling invasions of exotic species are an estimated €453-1036 million per year.
- the total economic losses are an estimated €665-1921 million yearly.

That adds up to a total cost of an estimated €1118-2957 million per year.

We further assume that most of the remaining *NA* items are fairly minor, or there would have been some indication of the costs. However, the costs of pest control and damage for the house longhorn beetle and zebra mussel amount to millions of euros per year. We estimate the total for all *NA* items to be €20-50 million annually.

Synthesis

These various methods produce highly divergent results. At this time, we do not place much value on extrapolations based on surface area, population size or gross national product because they lack a sturdy theoretical or empirical basis. There is still no consensus as to which species should and which should not be included in the calculations, and as to which costs should be included, and in what way. These questions require further study and agreements, nationally as well as internationally.

Awaiting further research, **the total cost from invasive exotic species to the Dutch economy is an estimated €1.1-3.1 billion per year. That equals 0.2-0.6% of the annual GDP.**

Costs by taxon

Like in the previous chapter, it is interesting to look at

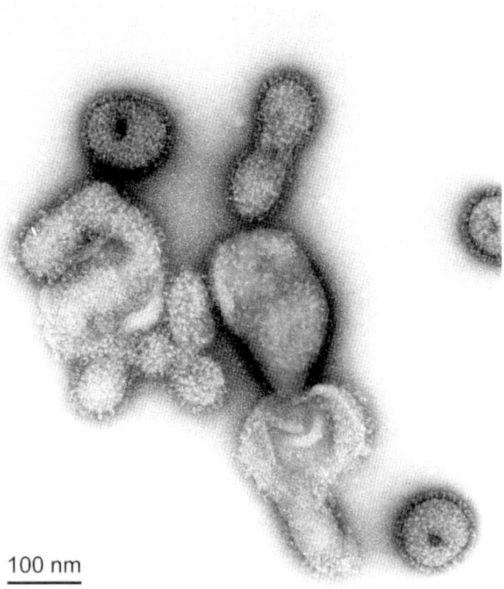

Of all invasive species in the Netherlands, the tiniest ones (viruses) generate the highest economic costs. Number 1 is the influenza virus.

such figures from a taxonomic point of view. That is more difficult in this case, since we have broken down the costs by economic sector rather than taxon. Even so, knowing that influenza, HIV/AIDS, swine fever, foot-and-mouth disease as well as bird flu are all caused by viruses, a quick glance at Table 10 makes clear that **viruses** are by far the main cost generator among the invaders in the Netherlands, accounting for at least 50% of total costs from invaders, even if we disregard the unknown but substantial costs from plant viruses. Paradoxically, the tiniest creatures generate the highest costs.

This picture differs from the US, where mammals and microbes (viruses *plus* bacteria, fungi and nematodes) dominate the scene. But as mentioned, the US figures disregard productivity losses from human diseases and overestimate the costs from mammals (feral cats). So in fact, microbes dominate the scene in the US. The high share of viruses in the Netherlands can perhaps be explained from the high densities of people as well as livestock.

It would be worth wile to make further analyses of the costs and losses from invaders by taxon, region of origin, and pathway.

The VOC and the WIC:
key players in globalisation

Inspired by the Portuguese and Spanish discoveries, conquests and trading voyages, entrepreneurs in the Dutch Republic in the late 16th century became aware of the enormous new trading opportunities offered by the East and West Indies. The first small merchant fleet left Amsterdam for East India in AD 1595, followed by several other voyages from Dutch ports. They benefited greatly from detailed information on the route, the coasts and peoples collected by Jan Huygen van Linschoten, who was secretary to the Portuguese archbishop of Goa from 1583 to 1589. After his return to Holland in 1592, transcriptions and printed copies of his *Itinerario* were keenly studied.

The VOC and globalisation

It soon became clear that extensive competition would prevent all the players from making large profits, so a trust was formed. In 1602 entrepreneurs and investors joined forces and capital to found the *Vereenigde Oost-Indische Compagnie* (VOC, Dutch East India Company). The purpose was to send cargo ships to the Far East for trade in exotic goods that could be sold at high prices on European markets. Depending on the return on investment, shareholders would then receive a dividend. This turned out to be a successful idea. The VOC is regarded as the first-ever multinational corporation, and the top multinational in the 17th and 18th centuries. For about 40

Ship of the VOC (Painting from William Fehr Collection. From: Insight Guide Zuid-Afrika 2002).

Biological globalisation

years in the 17th century its headquarters in Amsterdam were a kind of 'Oval Office' from where the VOC 'ruled the waves' and dominated world markets through its firm grip on local markets, prices and trade. Where it was deemed to be necessary, military force was used.

The VOC founded many trading settlements and forts along the coasts of South, Southeast and East Asia. From 1638 to 1799 the VOC was even the only foreign party allowed a trading settlement in Japan, founded on the island of Deshima near Nagasaki (a Dutch privilege that would continue until 1853). The VOC took over power in Sri Lanka, Taiwan (temporarily), Java and the Moluccas (Spice Islands), while simultaneously losing ground in Arabia and India. In total, the VOC organised nearly 5000 voyages to the Far East (or 8000 if return trips are included), an average of 26 voyages annually. Many ships did not return; some because they were shipwrecked, hijacked or dismantled, many others because they were kept in Asia to play a part in the lucrative trade *within Asia*.

The main money-makers were nutmeg, cloves, mace and cinnamon, of which the VOC managed to obtain and sustain trading monopolies. Other key goods were pepper, tea, coffee, sugar, textiles, silk, cotton, indigo, hides, cowry shells, silver, gold, tin, copper, sandalwood, saltpetre, Chinese porcelain and even opium. These goods were traded mainly for silver and gold (specie as well as bullion) and cloth, and occasionally for amber and African ivory.

In 1798, the VOC went bankrupt for various reasons, including poor governance, corruption among its overseas staff and a costly war with the British.

The WIC and globalisation

The *West-Indische Compagnie* (WIC, West India Company) was founded in 1621. It concentrated on trade between West Africa, South America, the Caribbean and the home country. The WIC founded many trading settlements and forts along the coasts of West Africa, Angola (briefly), Brazil (briefly), Surinam and the Lesser Antilles (mainly Curaçao). Military force was used many times to enforce and defend trading monopolies against Portuguese, British and French competitors, as well as private free riders, some of them Dutch.

The WIC did not organise as many voyages as the VOC. Between 1674 and 1740, 334 two-way or "triangle" voyages were organised, i.e. on average 5 per year. In addition, 383 ships were sent for the transatlantic slave trade. The WIC went bankrupt in 1740 due to loss of market share, costly sea wars and attacks on its monopolies from competing European countries as well as entrepreneurs in the home country. The forts were transferred to the Dutch government, which in 1874 sold the last ones in Ghana to the British.

The main money-makers for the WIC were gold, ivory and slaves, all from West Africa. Whereas the Portuguese and Spanish had started slave transports in the early 16th century, the WIC joined them in 1630 after their occupation of Brazil (which lasted until 1654), and particularly after 1641, following their occupation of the Angola coast (which lasted until 1648). The Dutch were the leading transatlantic slave traders during 1636-1672. In total, the WIC traded 280,000 slaves, or 2.5-3% of the whole transatlantic slave trade. In addition, small Dutch enterprises traded an almost equal number of slaves, bringing the total share of the Dutch in transatlantic slave trade at 5%.

Goods such as camlet, wood and wax were much less important and were traded mainly for textiles, both from the Netherlands and from South Asia (the latter purchased in the Netherlands from the VOC!). Guns, gunpowder, spirits and metals were also sold as well as cowry shells from the Maldive Islands, which were purchased from the VOC, too. At the time, these shells were used as a currency within West Africa.

Sources: Den Heijer (1997) and Gaastra (2002)

The VOC and the WIC as spreaders of species

Apart from being key players in world trade and globalisation, the VOC and the WIC were also important players in *biological* globalisation.

The VOC spreading species

The VOC introduced several crops into Indonesia, including:

- the coffee plant from Yemen;
- the cassava plant from Africa (where it previously had been imported from South America by Portuguese slavers);
- various vegetable crops from Europe (cultivated in the mountains of Java).

In addition, the VOC helped to spread maize, which had been introduced in the early 1500s by the Portuguese. In the 19th century, after the bankruptcy of the VOC in 1798, the Dutch also introduced the kina tree (producer of quinine, an anti-malaria drug) from South America.

The VOC hired scientists as early as the 17th century. One of them was the German Rumphius, who wrote two famous books on the plants and shells, respectively, of the Moluccas. His books suggest he personally introduced some plant species himself (P. Boomgaard, pers. comm.).

Medical botanists stationed in Deshima purchased, collected and shipped seeds an dried plants and seeds to Southeast Asia, Great Britain, Sweden and the Netherlands. Successful collectors were the German Engelbert Kämpfer and the Swede Carl Thunberg. In the 19th century, the Dutch hired another German, Philipp von Siebold. Among his numerous introductions were valuable Japanese tea varieties in Java. In Europe, often first in the Netherlands, he introduced many ornamental trees, shrubs and other species, including species and varieties of the genera *Lilium, Hydrangea, Prunus, Acer, Wisteria* and *Weigelia*, as well as *Iris ensata* (Carla Teune pers. comm.). A less valuable introduction was Japanese knotweed *Fallopia japonica*, which 150 years later, perhaps in response to climate change, began to develop into a noxious weed across Europe.

In South Africa's Cape Province, the VOC introduced Mediterranean citrus species and even, after many failures, the entire range of crops grown in Dutch vegetable gardens (see Box 13). In addition, the American prickly pear cactus *Opuntia ficus-indica* was introduced, probably from India. By 1750, this cactus had developed into a noxious weed, although it also provided food for the poor (see Box 40).

Live animals were traded as well. The VOC had a monopoly on the export of elephants from Sri Lanka. It also traded cattle from India to Indonesia and horses and goats, mainly from Arabia, to various Asian countries, including Japan. Almost inevitably, animal diseases were introduced, too. One example was the African horse disease (Afrikaans: *perdesiekte*) brought from the Cape Province to Indonesia and Europe.

As for biodiversity, habitat destruction combined with the introduction of pigs, goats, rats, dogs and monkeys had a devastating impact on island biotas. Among the victims was the famous endemic dodo *Raphus cuculatus* of Mauritius. It must have gone extinct around 1665. The VOC also contributes to the destruction of endemic flora and fauna on St. Helena (see Box 36).

We may safely assume that the trade within Asia and between Asia and Africa generated more species interchange than the trade between Asia/Africa and Europe, since Europe has a different climate. However, the *added* interchange by the VOC was limited by the fact that there had been trade and species interchange across the Indian Ocean since Greek and Roman times.

The WIC spreading species

The WIC may have contributed more to bio-globalisation than the VOC since transoceanic trade across the Atlantic started only after 1491. Furthermore, the biotas of the Old World and New World were much more different than the biotas of Asia and Africa.

A clear example is the transport of African slaves and European contracted labourers to America. The suffering of the slaves was only equalled by the suffering of the Native Americans felling victim to two African diseases introduced by the slave ships, malaria tropica and yellow fever. Whereas *Plasmodium falciparum*, which causes malaria tropica, had already been introduced in the early early 16th century by the Portuguese or the Spanish and their African slaves, we hypothesize that the Dutch are a likely suspect to have introduced yellow fever.

The evidence is circumstantial. The first documented outbreak was on Barbados in 1647. That was in a period (1636-1648) when the WIC dominated the transatlantic slave trade. In addition, there were small Dutch enterprises trading slaves. During the 1640s Dutch merchants had stimulated sugar cultivation in the Caribbean in order to create new trading opportunities. British landowners on Barbados were the first to start sugarcane plantations and needed slaves. In 1638 the WIC had started a slave market on Curaçao for sales to the Spanish, French and British colonies in the Caribbean. In addition, small Dutch enterprises supplied slaves directly to these colonies. Though the Dutch were the leading transatlantic slave traders they had lost the Brazilian slave market in 1645 due to a rebellion of the Portuguese population, so they

The dodo on Mauritius was driven to extinction by the Dutch and their introduced animals.

were seeking new markets for their slaves. In this context, it seems likely that it was Dutch slave traders who introduced yellow fever on Barbados, and thereby in the New World.

In addition to yellow fever, slave transports caused a repeated influx of several pathogens that had been introduced earlier from Europe and Africa (see Box 8). The European colonists themselves also suffered, mainly from malaria and yellow fever.* Even the black slave population was not entirely immune.

The WIC also traded in plant seeds and live animals, including civets *Viverra civetta* from Africa. They were valued in Europe for the oily substance they secrete, which was used in perfume manufacturing. However, we know of no civet invasions in the Netherlands, no doubt because of climate differences. Again, most species interchange must have taken place between South and South, viz. between Africa and South America/the Caribbean.

One of the admirable achievements of the VOC and the WIC was their extensive and precise documentation, mainly recorded in *daghregisters* (daily reports). Of the WIC reports, many have vanished, but of the VOC reports most have survived. Unesco has even entered the VOC archive on the World Heritage list. These archives, which fill kilometres of shelves, can be regarded as huge haystacks in which many needles providing clues about biological globalisation wait to be discovered.

* The second-generation European colonists in the Americas were also increasingly hit by Eurasian diseases, since they had not acquired immunity to these diseases.

Sources: Den Heijer (1997), Gaastra (2002) and McNeill (1998)
http://www.nda.agric.za/vetweb/History/H_Diseases/ H_Animal_Diseases_in%20SA7_htm

Introduced plants in New Zealand, the Cape Province and Amsterdam: how to explain their geographical origin?

The biotas of most regions in the world have a complex geographical origin. Most native species originate in the same region or in adjacent regions, some in more distant regions. Geographical barriers - both recent and old - also play a part. But where **introduced** species are concerned, distance and barriers are much less important.

A clear example is **New Zealand**. As for its native biotas, Europe shares very few species with its antipodes. New Zealand even lacks such cosmopolitan species as the peregrine falcon *Falco peregrinus* and the barn owl *Tyto alba*. However, in AD 1835 Darwin found many European plant and animal species invading the country (see Box 22). In 1860, the British botanist Hooker published a list of 61 plants naturalized in New Zealand. No less than 36 (59%) of these plants were from Europe (Crosby 2000).

This shows that whereas a long distance is an obstacle to species colonization and introduction (in fact, the Europeans themselves started their colonization only in the late 18th century, three centuries after 1492!), it is not an obstacle to introduction *success*. It also shows that plants from the Northern Hemisphere can thrive in the Southern Hemisphere, in spite of the contrasting seasonal rhythm.

Three factors help explain the geographical origins of exotics in New Zealand. New Zealand's **climate** resembles that of Northwest Europe; there was much **trade and traffic** with Britain; and as for **culture**, British colonists wished to introduce familiar ornamentals from their homeland (as well as animals, such as the pig and the blackbird). In 1839, a company that was trying to persuade Britons in migrate to New Zealand wrote (cited in Crosby 2000):

'In whatever part of either Island they have been planted, European vegetables, fruits, grasses, and many sorts of grain, flourishes remarkably, but no more than the different animals which have hitherto been imported, such as rabbits, goats, swine, sheep, cattle, and horses'.

A fourth relevant factor is the **diversity of species pools**. The Northern Hemisphere has a much larger landmass and hence a much richer species pool than has the Southern Hemisphere. This primarily holds for the tropics but to a lesser extent also for the non-tropical zones (Schilthuizen 2006). The southern non-tropical land mass only comprises Central and Southern Australia, New Zealand, the southern third of South America and species-poor Antarctica. This fact alone makes it more likely for *any* region in the world, including a temperate region such as New Zealand, to

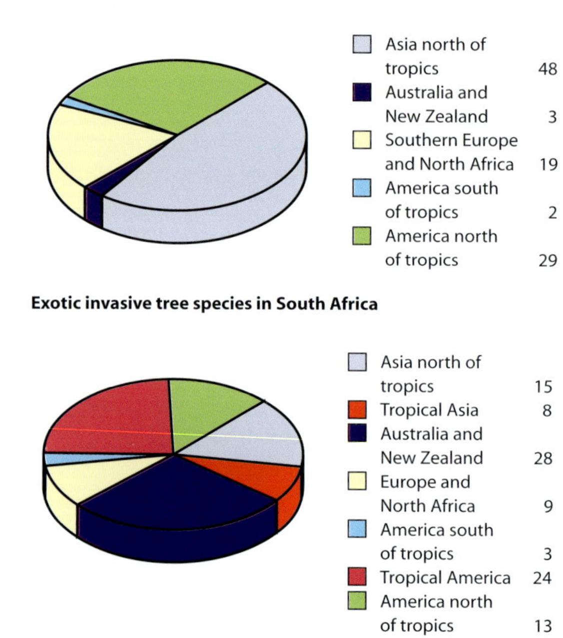

Figure 14.
Geographical origin of exotic tree species in southern Amsterdam compared with exotic invasive trees in South Africa.

obtain introduced plants and animals from the Northern Hemisphere than from the Southern Hemisphere.* This effect is reinforced by collectors. Searching for exotic plants they could introduce in their homeland, they often preferred regions with a matching climate and a high biodiversity. European collectors, for example, were eager to collect plants in Japan and China (see previous Box).

A fifth factor is the **competitiveness** of the species pool. Darwin was amazed at the one-sidedness of the exchange of life forms between Britain and New Zealand and in *The origin of species* stated his conclusion that:

> 'the productions of Great Britain stand much higher in the scale than those of New Zealand. Yet the most skillful naturalist, from an examination of the two countries, could not have foreseen this result'.

Even today, the Netherlands has just five naturalised plants from New Zealand (Van der Meijden pers. comm.). The New Zealand flora is less rich as well as less competitive than the flora of Western Europe, no doubt because of its long geographical isolation.

To illustrate the significance of these factors, we compare the origins of introduced trees of Amsterdam and introduced trees in South Africa.

Lever (2002) compiled an inventory of **tree species in the South district of Amsterdam.** In an area of just 10 km² he found no less than 170 tree species, most of which can be regarded as exotics: 58 species originate from Central, Western and/or Northern Europe, and 111 are native to Southern Europe or other continents.** A clear example of botanic globalisation. The 111 species break down in 19% from Southern Europe and/or North Africa, 48% from Asia north of the tropics, and 29% from America north of the tropics, leaving just 3% are from Australia and 2% from America south of the tropics.

Not a single tree from the tropics was found, so the climate factor again seems dominant. Nor is it surprising that species from the Northern Hemisphere dominate. This can be explained from the northern non-tropical land mass being much larger than the southern. But the northern dominance of 95% of the species is so overwhelming that other factors seem to play a part, particularly preferred trade routes. For example, Japan granted the Netherlands a trading monopoly from 1683 to 1853 and the Dutch hired various botanists who collected and shipped seeds and plants (see Boxes 72 and 73). This may be one of the reasons that of the 111 exotic trees in Amsterdam, 26 are native to Japan.

A different pattern is found in the trees of **South Africa**. An inventory by Henderson (2001) of **introduced tree species,** focusing on species that have become weeds and invaders, gives 95 species native to a different continent or from North Africa. These 95 species break down in 28% native to Australia and New Zealand, 8% to tropical Asia, 15% to Asia north of the tropics, 3% to America south of the tropics, 24% to tropical America, 13% to America north of the tropics and 9% to North Africa and Europe.

As expected on climatic grounds, South Africa has many more exotic trees from the tropics (32%) than has Amsterdam (zero). Nor is it surprising that among the non-tropical species, the northern species (37%) outnumber the southern (31%). But the northern dominance is much less than in the trees of Amsterdam (as well as the flora of New Zealand, although those figures date from 1860). So other factors must be involved. Again, preferred trade routes are a likely candidate. We may safely assume that South Africa has - and had - relatively more trade and traffic with Australia and New Zealand than has the Netherlands.

Of course, these analyses are rather straightforward. More detailed analyses will reveal closer relationships with the various factors mentioned as well as other relevant factors.

*   Logically speaking, the opposite pattern may be expected for marine species. However, we know of no empirical evidence supporting this hypothesis.

**   AD 1500 was chosen as a reference year for Amsterdam as well as South Africa, but some Southern European trees may have been introduced earlier in the Netherlands.

The mollusc that almost destroyed the Netherlands

In the 18th century, six of the "states" of the Netherlands, including Holland, Zeeland and Friesland, came under attack of an invasive species. The invader was the shipworm *Teredo navalis*.

The shipworm is not a worm but a mollusc that can grow up to 120 cm and bores tunnels in wood, including wood of ship's hulls, causing much damage. Its origins are unknown; it may have hitched a ride with ships sailing from Asia. From 1730 on, it caused heavy damage to ships as well as the wooden sluices and the wooden walls then used in most Dutch dikes. The artificially protected seacoast of the Netherlands stretched much longer at the time than today, because it included the borders of the inland sea Zuyderzee, which was not yet enclosed. Dikes broke during winter storms and the lower parts of the Netherlands were in jeopardy. Many prayer meetings were held.

Multiple efforts were made to ward off the danger. Ships and sluices were protected applying copper nails or even copper plates (copper being toxic to molluscs). In addition, the hulls of ships were doubled, whilst sluices were replaced every year. Protecting the dikes was much more difficult. It was decided to replace the wood with rock. Since this material was not locally available it had to be imported in high quantities from the Ardennes, Wales, Germany, Denmark and Norway. Costs amounted to 5 millions guilders in the north of North-Holland alone – an astronomically high figure at the time. Many farmers – who were the first to pay for water management – went bankrupt. For stone supply, even gravestones and chambered tombs (archaeological monuments from the eastern regions of the Netherlands, protected since 1734) were taken.

Although the key parts of the dikes could be protected within a few years, the operation took about two centuries to be fully completed. The expensive rock, and gravestone fragments, can still be found along the shoreline of what is now the IJsselmeer. They remind us of a country in panic.

The invader still survives along the North Sea coast, but, of course, its harm has been far reduced. Dikes along the seacoast are at the seaside covered with stone or concrete, the plating of most ships is made of steel or polyester and wooden sluices are being treated with creosote.

Sources: Den Hartog & Van der Velde (1987), Bieleman (1992) and Wolff (2005)

Shipworm burrowing in wood.

The Netherlands:
distribution country of ballast water

Since about 1960, mammoth tankers carrying huge amounts of ballast water are sailing fast across the oceans. These ships often carry dozens of species of organisms. What share does the Netherlands have in this transport?

Two-thirds of the ballast water arriving in the Netherlands stems from other countries around the North Sea. There is no major threat of bio-invasions in those cases. A much higher risk seems to be posed by ballast water from the east coast of North America, which happens to represent a large proportion of the ballast water arriving in the Netherlands.

Furthermore, there are modest, though not risk-free loads from:
- Central America
- South America
- Asia (particularly Japan)
- South Australia
- New Zealand
- Oceans
- the (western) Mediterranean
- the Azores and Canary Islands.

Some ballast tanks contain water from multiple regions.

One study of ballast tanks found 170 species of phytoplankton and zooplankton, including three exotic species, all dinoflagellates (Wetsteijn & Vink 2001). In addition, in some of the ballast tanks some (potentially) toxic phytoplankton species were found.

Another study compared the import and export of ballast water. The result:
- Some 7.5 million tons of ballast water are dumped in Dutch waters annually.
- Much more ballast water is taken in: some 70 million tons.

In other words, the Netherlands is a major net exporter of ballast water. The destinations of this water remain unclear, since ships leaving the Netherlands often stop at multiple foreign ports.

Though small compared to the export volume, the import is yet significant if compared to other European countries: 42% of the total ends up in the Netherlands! Clearly, the country serves as a distribution country for ballast water.

Looking ahead, with Rotterdam as the world's third largest seaport only after Shanghai and Singapore, and with Singapore having a different climate, the Shanghai-Rotterdam route may well become the most important two-way avenue for bio-invasions.

Sources: Aquasense (1998a) and Wetsteyn & Vink (2001)

The Japanese oyster in the Netherlands: a derailed introduction

The Japanese oyster *Crassostrea gigas* was introduced in the Netherlands in 1995 to compensate for the collapse of the flat oyster *Ostrea edulis* that had followed the harsh winter of 1962/63. Experts expected that the Japanese species could not reproduce in the Dutch climate. In addition, it was expected that the Oosterschelde (East Scheldt) estuary would be closed off and become a freshwater basin.

Both assumptions proved wrong. Larvae of the Japanese oyster escaped from aquaculture farms and settled in the estuary on a massive scale. By the year 2000, the Japanese oyster was already covering 12% of the available area for shellfish in the Oosterschelde, and this percentage is expected to increase much further (A. Smaal pers. comm). Meanwhile, the oyster has also expanded into Lake Grevelingen and the Wadden Sea.

The Japanese oyster swamps on many locations. It covers other species, takes up the available space, competes with other filterfeeders for food and even feeds on their larvae. It thus displaces not only the flat oyster but also mussels and cockles. By 1999, the number of cockles in the Oosterschelde had already dropped by 70%. Cockle fishing and mussel farming are suffering damage, though no specific monetary figure is available. In the fishery sector, estimates range from zero to several million euros annually (J. Craeymeersch pers. comm.).

In addition, there are ecological effects. Ducks and waders are not able to forage on the oversized Japanese oysters. On the other hand, the oyster forms a hard substrate on which an entirely new biocoenosis can settle.

The invader also impedes recreational activities. Beaches can no longer be used for swimming and wading because the sea floor is covered with sharp and hard oyster shells.

Efforts to control the oyster are still small-scale. Some attempts are being made to clear away oysters, particularly on popular beaches. As yet, the cost of this is marginal.

Japanese oyster swamping.

Pumpkinseed sunfish driving out indigenous fauna in Dutch ponds

Ecological damage from an invader sometimes becomes apparent only decades after introduction. A clear example is the pumpkinseed sunfish *Lepomis gibbosus* in the Netherlands.

This fish was introduced some 115 years ago from North America. In the 19th century, many fish species were imported to Europe in order to test their use as a food fish. The pumpkinseed sunfish was brought in as feed for North American pike species. The pikes did not survive the journey but the sunfish did, and was then released into various bodies of waters. As long as there were sufficient predatory fish (pike, perch) the number of pumpkinseed sunfish remained limited.

When the garden pond came into fashion, and particularly since the 1980s, the beautifully coloured fish got a second chance. Since it reproduces rather quickly, it often caused overpopulation of the aquarium or pond after a few years. Owners released surplus fish into natural bodies of water, including small-sized basins like natural pools and ponds. Here, the same happened as in the aquariums and gardens ponds: in a short period of time, the populations had grown to huge numbers. In pools, the pumpkinseed sunfish has little to fear from predation and competition, because there often are no other fish species.

The pumpkinseed sunfish, which grows to a length of up to 15 centimetres, is a predator feeding on just about anything that passes by and isn't too large. It prefers small crustaceans and insects, but also happily feeds on amphibians and their larvae. The numbers of indigenous fauna are often decimated by the fish. In many locations, Red List species have disappeared, such as the internationally threatened Italian crested newt *Triturus carnifex*, the extremely rare garlic toad (common spade foot) *Pelobates fuscus*, and the dragon species *Lestes virens* and *Ceriagrion tenellum* (small red damselfly). Because the fish also feeds on zooplankton, the ecological balance in such bodies of water can be severely disrupted.

Efforts to combat the pumpkinseed sunfish are often ineffective. One of the few effective methods is to repeatedly pump out all the water. Obviously, though, this method is also fatal to other organisms.

The trade in these animals is no longer acceptable. Traders as well as private aquarium keepers should be made aware that releasing pumpkinseed sunfish into natural bodies of water may well cause severe damage to the indigenous fauna, including some Red List species.

Source: Van Kleef et al. (in prep.) and De Nie (1997)

The larvae of the common spadefoot *(below)*, a red list species, are eaten by the pumpkinseed sunfish *(left)*.

Economic costs and benefits from introduced aquatic species in the Netherlands

Aquatic exotic species can cause benefits as well as costs for the economy.

The *potential* economic damage from organisms in **ballast water** in the Netherlands was estimated by Broersen *et al.* (Aquasense 1998b) at:

- €12 million for fishery (= 5% of the total production value);
- €42 million for the shellfish industry (= 90%).

The latter are the costs of a whole year's suspension of activities related to the occurrence of plankton-induced shellfish diseases such as PSP and DSP (Note: these are human diseases; the mussels themselves are not affected). In reality, these activities are more likely to be suspended for several weeks or months. We therefore use a figure of 10% of €42 million = €4.2 million. That brings the total to €12 + €4.2 = €16.2 million. Assuming this damage occurs every three to five years, as was found in Sweden, this amounts to an average of once every four years. This yields an *average damage of €4 million per year.**

Myticola intestinalis, a copepod parasite of oysters, was considered a serious danger after its discovery in 1948. Mussel farming was even reorganised in response. Today, the species is considered relatively harmless (Engelsman & Haenen 2003). The species occurs commonly in the Netherlands. There are also other *Myticola* species parasitising oysters.

The parasite ***Bonamia ostreae*** arrived in the Netherlands through oyster transports. Flat-oyster farming never managed to recover from the extremely cold winter of 1962/63, something for which *Bonamia ostreae* can be held partially responsible. This paved the way for the introduction of the next species:

The **Japanese oyster *Crassostrea gigas*** is being harvested and traded for consumption. It heavily competes with mussels and cockles for food and space. There are 20 companies dealing in Japanese oysters. Turnover amounts to €5 million, and employment amounts to 60 man-years annually.

Estimated profit of oyster fisheries: € 0.8 million per year, 90% of which can be attributed to the Japanese oyster. Estimated damage to fisheries: approximately €1 million annually.

So even if we leave aside the unknown damage to recreation, it appears that the damage from the Japanese oyster exceeds the profit.

The **Chinese mitten crab *Eriocheir sinensis*** damages fishing-nets, but it also appears on restaurant menus. To our knowledge, there are no available figures on the costs and benefits from the crab.

The **shipworm *Teredo navalis*** is not a worm but a mollusc. The shipworm's origins are unknown; it may have hitched a ride with ships sailing from Asia in the 18th century. The shipworm damaged sea-dikes and generated very high costs in the 18th and 19th centuries (see Box 75).

Today, the **muskrat *Ondatra zibethicus*** undermines inland dikes at a large scale.

Costs: € 23 million per year, not counting labour costs for water boards and repair costs.

Large seaweeds like **japweed (Japanese wireweed)** ***Sargassum muticum*** can block ships' propellers and clog up harbours. The costs of this are not known, but would not appear to be very high.

Locally, **Carolina fanwort *Cabomba caroliniana*** is

The Chinese mitten crab damages fishing-nets.

swamping and some water boards suspect that **water primrose** *Ludwigia grandiflora* and **parrot's feather** *Myriophyllum aquaticum* may become invasive and clog watercourses.

Already a real pest is **water pennywort** *Hydrocotyle ranunculoides*. This native to North America was imported as an ornamental. In 1994 it was found for the first time in open water, where it proliferated fast. It must now be periodically combated at many locations. This is done manually as well as mechanically and - where those methods are insufficient - also chemically, using MCPA. Control is difficult because the plant can reproduce and spread through tiny cuttings. One *waterschap* (water board) having a big water pennywort problem is De Maaskant, which oversees polders west of the city of Den Bosch, between the river Meuse and a canal named the Drongelse Kanaal. Propagule pressure is extremely high due to the continuous supply of plant cuttings from the canal. The cuttings are produced by mowing in a neighbouring water board. Costs are high, since the plants accumulate heavy metals and hence must be treated as chemical waste after removal (the positive effect of this, of course, is that the water becomes cleaner).

In water board De Maaskant (9000 hectares) the control costs are €120,000- 140,000 per year (R. Pot pers. comm.).

In the region of water board Stichtse Rijnlanden, the costs of a one-time "spring-cleaning" are €840,000 for a subcontracted company plus 740 hours for their own staff. Assuming the gross salary cost is €50 per hour, costs amount €877,000. If we further assume the frequency of the "spring-cleaning" to be once every five years, the average annual costs in Stichtse Rijnlanden amount to €175,000. In addition, permanent monitoring, public information and local cleanups (at an annual cost of €20,000) are needed.

Based on this, we can provide a high and a low estimate for the Netherlands at large. Our low estimate equals 6 times the combined costs for Stichtse Rijnlanden and De Maaskant; and our high estimate is 12

Water pennywort swamping.

times the joint costs for these two areas.
Estimated total costs for the Netherlands: 6-12 x (€ 130,000 + € 195,000) = € 2- 4 million per year.

* These figures are based on a rather straightforward argumentation. The reality often proves more complex because ballast water can be taken in and let out more than once. In addition, many different kinds of organisms and cysts can survive in the silt at the bottom of the ballast tank. The tank can contain tens of centimetres of silt.

Costs from livestock pathogen invasions

Rinderpest, plagued the Netherlands for two centuries. This intestinal viral disease presumably originated in South Asia. It first struck the Netherlands in 1713. There were two additional waves in that same century. The one in 1744 was a complete disaster: it killed no less than two-thirds of the cattle in Holland and Friesland (Frisia). Between 1845 and 1848 there again was high mortality following the import of infected cattle from the Baltic States. In 1865, 150,000 cows fell ill and 27,000 were culled.

Scrapie is a degenerative sheep disease caused by a prion. It has struck British sheep since the beginning of the Industrial Revolution. Previously it occurred on Iceland, from where it may have spread to the British Isles with sheep transports associated with the export of small Icelandic horses for use in mines. In the Netherlands, scrapie was first diagnosed in the 1950s. Today it occurs on 6% of the sheep farms. Since scrapie is related to the zoonosis BSE, pressure is being put on the sheep farmers to eliminate the disease. The main control strategy is to eliminate the most susceptible genotypes in breeding programmes.

BSE (Bovine Spongiform Encephalopathy) or "mad cow disease" is another degenerative disease caused by a prion. The disease may have evolved from scrapie. The first BSE cases in the Netherlands probably originated in the UK (Prusiner 1996, Bol 2001). The costs of the disease were high. The EAAP (2003) calculated the combined costs for all EU countries at €2.8 billion per year. The loss of income for industries was expected to total €1.5 billion per year. In the Netherlands alone, the annual costs of prevention and control were estimated at €519-769 million, and the damage at €35-171 million (Appendix 3).

Foot-and-mouth disease is a virus disease of cattle that struck livestock In Europe many times, at least from AD 1514, but perhaps much earlier (Hubbert *et al.* 1975). The virus has come and gone many times. It can also hit humans. The most recent outbreak in the Netherlands was in 2001 – a secondary invasion from Africa via the UK (see Box 21). The virus raged from March 14th until April 22nd. Infected areas were stringently closed off and large-scale animal culling took place. In total, the epizootic cost an estimated €874 million (Appendix 3).

Classical swine fever is another virus disease. It was by far the most costly invasion in the Netherlands of the last decades. It struck the country in 1997-1998, introduced by a contaminated car truck returning from Germany. It also hit parts of Germany, Belgium, Spain and Italy. In August 2000, southern England was also struck, after a pig infected by a wild boar was imported from Asia. Total estimated costs were between €1.5 and €1.9 billion, approximately 0.5% of the annual GDP (Appendix 3).

Bird flu (avian influenza) raged in the Netherlands from February 28th to August 22nd, 2003. The virus (type H7N7) may have arrived earlier. It is not known where it came from. The country's two most important poultry regions, the Veluwe and the Peel, were both hit. Costs were estimated at €870-970 million (Appendix 3).

Just as with swine fever and foot-and-mouth disease, the large-scale animal culling and destruction led to a public outcry, particularly because this time animals kept by non-farmers were culled as well.

Epizootics generate economic losers and winners

It is not really possible to speak of "the" costs to "the" society of a veterinary disease outbreak. First, the costs are very unequally distributed among those involved; second, there are also benefits; and third, the benefits are unequally distributed as well.

We can illustrate this with the outbreak of classical swine fever in the provinces of Brabant and Limburg in 1997:

- For the pig farms, slaughterhouses and livestock transportation companies involved, there is significant damage, not only due to loss of sales, but also because the companies need time to resume production after the transport ban is lifted.
- The costs of animal culling and destruction, and the damage to business are shared by farmers, the national government, the EU and the entire production chain, all parties suffering damage.
- Pig farmers in other countries and other Dutch regions may temporarily benefit from higher prices and can, perhaps permanently, expand their market share.
- If the EU allows vaccination, it will allow that only to animals produced for the domestic pig meat market. The question then remains whether supermarkets will purchase such meat. If not, slaughterhouses will not be able to sell the meat, and the livestock industry will bear the cost of vaccination without any benefit.
- This does not apply if consumers switch to beef and poultry meat; those industries are then able to benefit.
- If consumers start eating less meat generally, the entire livestock and meat industry will suffer damage.
- In that case, consumers will spend more money on other items and other sectors will be able to benefit.
- Animal destruction companies earn a large amount of money from animal destruction.
- Transport bans during the epizootic can also affect other industries, including hotels and restaurants.
- The threat of sudden border closings looms large over livestock farmers like the sword of Damocles. Insecurity over how supermarkets will respond adds to this. That, in turn, may affect willingness to invest, bank guarantees and insurance premiums.

In short, economically speaking, an epizootic – like any disaster – has losers as well as winners. The distribution of costs and benefits depends to a large extent on the policy. If the livestock industry would be held responsible for 90% or 100% of the costs of epizootics, a large part of the industry would not survive a severe outbreak.

Uninvited guests from abroad

Cockroaches have been indigenous in Europe for a long time, but some of the many species came from other continents during the 20th century, usually hitching a ride with cargo shipments. These include **Periplaneta (!) americana, Periplaneta australa, Pycnocelus surinamensis** and **Blatta orientalis** – their names revealing their presumed region of origin. They need warmth, and therefore like to settle near central heating and in kitchens, bakeries and greenhouses. Cockroaches are incredibly tough animals. For instance, they are serious candidates to survive a nuclear war, since they can survive nuclear radiation at levels hundreds of times higher than humans can. Even when decapitated an individual can survive for up to nine days, after which it starves to death. In greenhouses, the parasitic wasp *Aprostotecus hegenowii* has been successfully applied as a control

House longhorn beetle.

Cockroach.

method; it deposits its eggs in the cockroach's eggs.

In the Netherlands, the cockroach is the largest cause of asthma, second only to dust mites. This fact alone is responsible for a high cost. At least half a million people have asthma; the costs of treatment are €300-1000 per patient per year. Assuming the share of cockroaches in asthma to be between 10% and 20% and the average costs per patient to be €600 per year, the total costs amount to €30-60 million per year, not counting the costs of pest control. Lacking specific figures, we assume exotic cockroaches to be responsible for 50% of the costs: €15-30 million per year.

The **house longhorn beetle** *Hylotrupus bajulus* is a guest from the tropics, which destroys large amounts of wood. The pest is controlled by treating wood with chemicals and gas. The costs are substantial but not quantified.

The **house cricket** *Acheta domestica* is native to Southeast Asia and thrives in warm environments like homes, bakeries and greenhouses. It can hitch a ride with plants and fireplace wood, among others. It's chirping can be a nuisance. Poison is used for pest control.

The **pharaoh ant** *Momorium pharaonis* owes its name to the (probably false) assumption that it is Egyptian in origin. In fact, it originates in the tropics, where it has gained a reputation as a decomposer of cadavers. This fact has inspired biologists to develop suc-

Larva of house longhorn beetle.

Brown dog tick, engorged female.

cessful pest control methods using tubes containing meat, pheromones and poison. Worker females then transport the poison to their queens. Colonies count up to 300,000 individuals, including 30 to 400 queens. In the Netherlands, the ant was first reported in 1900 in a post office in the city of Leeuwarden, where it had probably arrived by mail. It thrives in buildings and homes with central heating, the preferred temperate being 30°C. From 1950 onwards, houses in the Netherlands became warmer. That coincided with the career of the pharao ant.

The **gray silverfish *Ctenolepisma longicaudatum***, a member of the silverfish family, is widespread in the tropics and subtropics of the Old and New World. In 1914 it was first found in Southern Europe. The first observation in the Netherlands was made in 1989 but the species probably arrived earlier with cargo from the (sub)tropics. This wingless insect lives in houses and other buildings, feeding on cellulose (mainly paper) and starch (including glue). It can damage books, other paper materials and insect collections (Beijne Nierop & Hakbijl 2002).

The **brown dog tick *Rhipicephalus sanguineus*** is 0.8 mm long, barely visible with the naked eye. First described in 1806, it is thought to have originated in Africa and was first observed in the Netherlands in 1962. By now, it has established, though it can probably only survive indoor. It attaches itself to the armpits and groins of dogs, its preferred host, as they walk through high grass or low scrub. However, cattle, cats and sheep can also also serve as host, as do humans. In some rare cases, dogs and cats brought from Southern Europe can lead to human tick bites, carrying a small risk of Lyme disease and – consequently – meningitis and arthritis. The ticks can also transmit rickettsias causing tick-borne encephalitis.

The **skin beetle *Dermestes haemorrhoidalis*** was introduced by shipping from South America in 1977. The beetle feeds on grain products and even zoological collections. One of its natural enemies, the parasitoid wasp *Laelius pedatus*, had been found two years earlier.

Costs from the influenza virus in the Netherlands

The near-annual invasion by one or several new strains of the influenza virus claims several hundred fatalities in the Netherlands every year. A much larger number of people are taken ill. What are the associated economic costs?

Using the same method as in Box 67, but adjusting assumptions, our low estimate can be €391 million per year; our high estimate amounts to €782 million per year (see Appendix 4).

The number of cases can be reduced by vaccination, and 72% of the target group does actually take the vaccine. This group consists largely of elderly people, so we can safely assume that only a small proportion of productive people are vaccinated. However, more employers are starting to offer free flu vaccination to their employees.

Flu vaccination financed by public funds costs about 31 million per year and results in a 53% reduction in excess mortality. Nevertheless, average excess mortality due to influenza is still counted in hundreds of people each year. From a purely economic perspective, the €31 million invested provides three benefits: a reduction in mortality of skilled employees; a reduction in sick leave; and a reduction in healthcare costs.

In addition to – and in spite of – these prevention costs, flu generates treatment costs. Unfortunately, the costs of treating infectious diseases are available only as a lump sum, not separated into categories. Instead, we make three assumptions: on average 1000–4000 patients per year are in hospital for one week at €1000 per day. This adds up to €7–28 million annually.

The total estimated minimum costs then add up to €391 + 31 + 7 = €329 million; the estimated maximum costs amount to €782 + 31 + 28 = €841 million. **The influenza virus costs the Dutch economy an estimated €329–841 million per year.** Of this, only €38-59 million can be counted as prevention and control costs, the lion's share is due to losses.

The costs from the flu would go sky high if the present H5N1 virus mutates into a highly virulent and contagious lineage and causes a **severe pandemic**. In that case, costs may amount to **€22 billion. Assuming a GDP of €448 billion per year (2005), that would make 5% of GDP**, well above the 3% estimated by the World Bank for the global economy. The open Dutch economy will be hit relatively hard by the collapse of international traffic and trade that is expected to follow a pandemic. The economy will shrink, and recovery may be slow, particularly if the age structure of the population is disturbed. Paradoxically, recovery may be the more slow since the virus will cause no material damage. After most natural disasters, economic recovery is enhanced by the challenge to repair material damage. That effect will not occur.

Sources: NYFER (2002), RIVM (2003), www.rivm.nl, www.kostenvanziekten.nl version 1.0 (June 2006)
Source for costs of vaccination: Zorgkompas (2003)
Source for bird flu: Kwartaalbericht De Nederlandse Bank (2006)

Costs from the HIV virus in the Netherlands

The HIV virus must have been introduced in the Netherlands in or some years before 1981. In that year, AIDS was first diagnosed in a man who had presumably been infected before 1980 in the US. By 1990, the number of fatalities had risen to some 450 per year, with an annual influx of about 450 patients. Mortality then dropped to 114 in 2003, in spite of an increased annual influx of 600 HIV-infected patients. In 2005 only 25 patients died.

This drop in mortality was due to the introduction from 1996 of combination therapy, or Highly Active Anti Retroviral Therapy (HAART). The costs of HAART increased until 2000 but then stabilised at about 9600 per patient annually, excluding the costs of clinical and polyclinic (outpatient) care. The total costs of HAART plus care are now an estimated 13,000 per patient per year.

In the second half of 1996, the number of patients were treated with HAART was 1688, rising to 9594 by the end of 2005. For the total number of patients on HAART, the costs amountled to 9594 x €13,000 = €125 million per year, which equals 0.2% of the total national healthcare budget of €57.5 billion in 2005. In addition, there were 2500 HIV-positive people not on HAART and they generate costs as well.

Further costs are incurred for prevention, including polyclinics for sexually transmitted diseases (STD), blood testing for HIV antibodies, consultations with doctors, public information, monitoring and research. We estimate these costs at €10 million (non-HAART patients) and €15 million (prevention) per year, making €25 million in total. This brings the total medical costs up to €150 million per year.

However, there are also much lower estimates. The National Institute of Public Health and Environmental Protection (RIVM), using a different calculation method, estimates the costs of HIV-related healthcare at only €20 million per year. Their calculation method tends to underestimate the costs, particularly the costs of clinical and polyclinic care (J. Polder pers. comm.). On the other hand, methods based on the number of patients multiplied by treatment costs tend to overestimate costs. Nevertheless, the study of Bogaards *et al.* (2004) seems relatively well documented. Therefore we use their estimate of €125 million as our upper limit and €50 million as our lower limit.

In addition to these medical costs, the virus generates prevention costs, mainly from increased use of condoms. However, these costs cannot be attributed entirely to HIV, since they will also prevent other STDs and reduce healthcare costs from these diseases. Hence we have not included prevention costs in our calculations.

Much more important are the indirect economic costs from productivity losses. Here we have to include a special feature of the high survival rate due to HAART: many patients are able to resume work, but almost exclusively on a part-time basis – often three days per week or less. Applying the same calculation method as used in Boxes 65 and 83, and adjusting assumptions, we have made calculations of productivity losses in Appendix 4. Our estimate is €53–160 million per year.

As for the total costs from the virus, our low estimate is €50 + 53 = €103 million per year. A high estimate amounts to €150 + 160 = €310 million. **The HIV virus costs the Netherlands an estimated €103–310 million per year.** These costs will rise as the influx of HIV-patients continues.

Sources: Bogaards et al. *(2004), HIV-Monitoring Foundation (2006) and www.kostenvanziekten.nl*

Malaria returning to the Netherlands: false alarm?

Malaria used to be endemic in the Netherlands. The last major outbreak, which was particularly heavy in the province of North-Holland, occurred shortly after World War II. By 1970, however, the WHO declared the country free of malaria. But as malaria expands its range across the world, driven in part by climate change, what is the chance of malaria reinvading the Netherlands?

Consider first the multiple factors involved in the spread of the *Anopheles* mosquito:

- The Netherlands has two mosquito species able to transmit the malaria parasite: *Anopheles maculipennis messeae* and *A. m. atroparvus*. In the past, there were two malaria parasites: *Plasmodium vivax* and the less common *P. malariae.*
- *Anopheles m. atroparvus* is associated with brackish water in coastal areas in the north and northwest of the country. It prefers to blood-feed on pigs and cattle rather than humans, and hibernates in livestock stables, from where it occasionally flies into adjacent homes to feed on humans. Since 1920, the species has become rare as farmers moved to homes not attached to the stables, as they were in traditional farmhouses. In addition, brackish waters have become less common after the construction of large dikes and dams. Today the species is very rare.
- *Anopheles m. messeae* still survives in low densities in association with freshwater, feeding on roe deer and rodents. However, it does not feed on humans.
- The Netherlands is still much too cold for tropical and subtropical mosquito species.

What about the parasite?

- Those *Plasmodium* species that affect humans have no other hosts as reservoirs, an ideal condition for eradication. Whereas *P. malariae* had become extinct after 1920, *P. vivax* could only be eradicated between 1945 and 1955 through a campaign of patient treatments combined with mosquito control using DDT. Water pollution further reduced the mosquito's survival opportunities. The last patient in the Netherlands was reported in 1961 and, lacking a reservoir, the parasite could not survive. Today, at any point in time (point prevalence) there are only a handful of malaria patients (import cases) in the entire country. Whereas the percentage of infected mosquitoes is low even in the tropics (often between 3% and 7%), the number of infected mosquitoes is virtually zero in the Netherlands. Hence, even for a person bitten by a malaria mosquito, the chances of infection are negligible.
- Of the several hundred reported cases of imported malaria, many were not caused by *P. vivax*, but by other plasmodia normally transmitted by *Anopheles* species that are unable to survive in the present Dutch climate. Yet, there were three cases in the 1980s in the proximity of Schiphol (Amsterdam Airport).
- A malaria patient has a high fever and is usually hospitalised soon after the disease is diagnosed. In the short period before entering hospital the chance of the patient being bitten by a (rare) *Anopheles* mosquito is minimal, let alone the chance that the patient will have developed parasites to the gametocyte stage, which is the only stage of the parasite that can infect mosquitoes; and the chances of that mosquito being infected and subsequently transmitting the parasite to a second person are even smaller.

So far, there seems to be little reason for concern. Even if *A. m. atroparvus* increases in numbers again, malaria could only return if a reservoir of *Plasmodium* in humans develops as well.

Of greater concern are two cases of autochthonous malaria tropica – which is fatal more often than other forms of malaria – reported in a German hospital in 2001. The parasite, *Plasmodium falciparum*, had probably been brought in by a child from Angola, but was transmitted by the indigenous (!) *Anopheles plumbeus*. As this mosquito occurs locally in the Netherlands, too, a competent vector for *P. falciparum* is already present in the country. *A. plumbeus* normally hides in hollow trees, but is broadening its habitat to include gutters and sewerage, perhaps in response to climate change. However, this species, too, can only become a serious threat if *P. falciparum* can establish a reservoir in humans. The key preventive strategy is adequate healthcare for every citizen, regardless of his or her legal status.

Sources: Van Seventer (1969), Takken et al. (1999), Takken et al. (2002), Knols (in e-mail), Kruger et al. (2001) and Bol (2002)

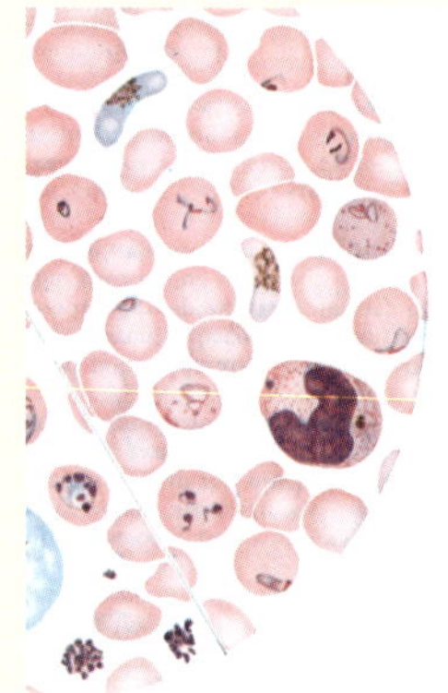

The malaria parasite *Plasmodium falciparum* in various stages of consuming red blood cells (From: Mackie *et al.* 1954).

More serious threats:
bluetongue, West Nile virus and *Aedes albopictus*

There are three more serious threats of pathogens invading the Netherlands than malaria, all of them insect-borne viruses.

Bluetongue

Bluetongue is an insect-borne viral disease that affects ruminants but not humans. Sheep are most severely affected, although cattle are the main mammalian reservoir. Bluetongue was first described in South Africa but has since been recognised in most countries in the tropics and subtropics. Since 1999 there have been widespread outbreaks in Greece, Italy, Corsica, the Balearic Islands and the Balkans, all located well north and west of its normal distribution. It appears that the virus has spread from both Turkey and North Africa.

The bluetongue virus (BTV) is an orbivirus transmitted by several species of biting midges of the genus *Culicoides*. The midges prefer cattle but turn to sheep when cattle have become absent or scarce. The disease cannot be transmitted from one sheep to another, except via the sperm and placenta. Symptoms include fever, oedema of the head, drooling and occasionally a cyanotic blue tongue; mortality may reach 70% of the infected animals. During a recent epizootic in the US, 180,000 sheep died in four months. Vaccines for sheep are available but not for every type of the virus and have adverse side effects.

In August 2006, the virus was found in the Netherlands and adjacent regions of Belgium and Germany, no doubt introduced with infected midges or cattle and probably facilitated by a hot summer. A year before, competent vectors were already found but it is too early to conclude that the virus is there to stay.

West Nile virus

Another serious threat is the West Nile virus, a flavivirus that originated in Africa, India and the Middle East. It usually causes a mild feverish condition, but 1% of patients develop encephalitis or meningitis, which is often fatal. West Nile virus is transmitted by *Culex* mosquitoes to birds and mammals, including humans. Infected mosquitoes have been observed in Southern Europe and the virus has already caused mortality among birds in Vienna. There is a good chance the mosquito will spread northward (partly due to climate change) and reach the Netherlands. The virus might already occur in migratory birds.

In 1999, mosquitoes or birds infected with the virus reached North America (New York City) killing 7 people. From there it spread to 29 states. By 22 August 2006 it had already caused 20,287 reported illnesses and 685 deaths among humans, and over 20,000 fatalities among horses. West Nile is now the dominant vector-borne disease in the US and Canada. Two vaccines for horses have already gained approval but, as yet, no vaccine for humans is available.

The risk to humans and horses in Europe appears to be smaller than in America, since the preferred host of the species in America is a migratory bird, the American robin *Turdus migratorius*. By the time this bird migrates south in late summer, the mosquito shifts to other hosts, including humans. In Europe, the host birds do not

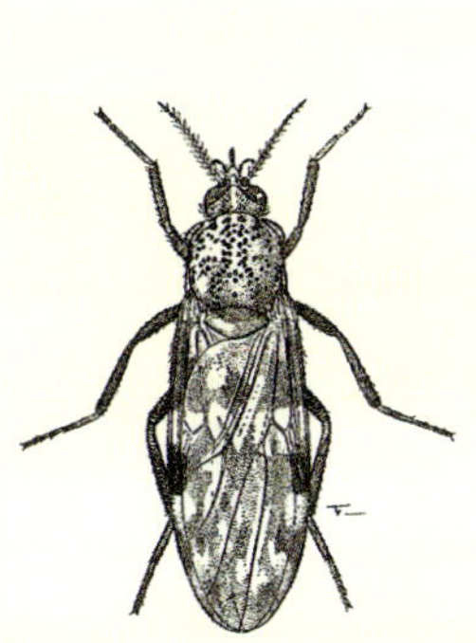

Culicoides (From: Smart *et al.* 1948).

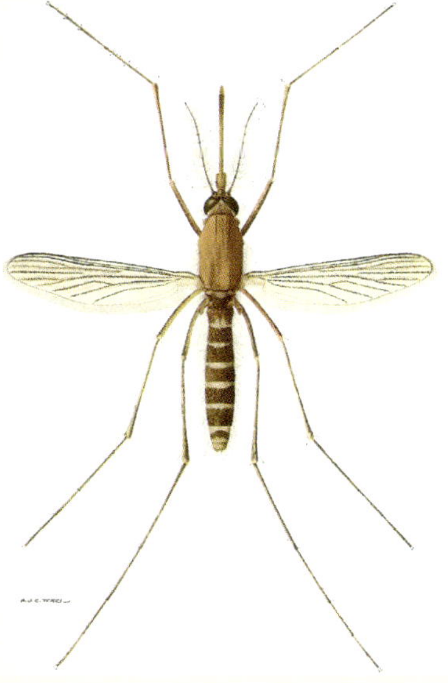

Culex (From: A.J.E. Terzi [from Edwards]).

migrate, so the mosquito can remain with its host. But of course, this behaviour may change over time.

Aedes albopictus

The Asian tiger mosquito *Aedes albopictus* is a subtropical mosquito native to Asia, from where it has spread to Southern Europe (1979), North America (1985), South America (1986), Africa (1991) and islands in the Indian and Pacific Oceans. In Italy it is now a dominant pest mosquito. The mosquito breeds in containers with standing water, such as scrap tires and tins, with which it is often transferred, even overseas (see Box 19). It has a wide host range among domestic and wild animals, including humans, and can transmit about 30 viruses, some pathogenic to humans: dengue, eastern equine encephalitis, Cache Valley virus disease and chikungunyavirus disease.

Aedes albopictus has spread widely in the eastern US. However, it is not a major threat to public health since it transmits fewer viruses than *Ae. aegypti*, with which it successfully competes. In Brazil, however, it is feared that it plays a role in the transmission of yellow fever from rural cycles to urban cycles sustained by *Ae. aegypti*. In islands in the Indian Ocean it has recently caused major damage. On La Réunion alone, which has 770,000 inhabitants, the mosquito between March 2005 and March 2006 infected no less than 212,000 people with the chikungunya virus, of which 148 died. Approximately 200 import cases have been reported in Europe.

There is a risk that the mosquito will invade the Netherlands in response to climate change. Given that in North America the mosquito ranges as far north as Chicago, this risk does not seem to be negligible. Locally, the mosquito has already been found in the Netherlands, near the Aalsmeer flower auction at companies trading

Dracaena or 'lucky bamboo' plants placed in water. It was probably imported on ships from southeast China, where dengue is endemic. Experts have warned that the mosquito can transmit the virus to the Netherlands and have advocated an import ban on aquatic plants from China, a measure already taken by the US and Canada. An alternative option is being considered: placing the plants in sterilised water or gel.

As put forward repeatedly in this book, most climate-driven bio-invasions cannot be halted but many trade-related invasions can be.

Sources for bluetongue:
www.addl.purdue.edu/newsletters/2002/spring/bluetongue.shtml
www.defra.gov.uk/animalh/diseases/notifiable/disease/bluetongue.htm

Sources for West Nile virus:
Spielman & D'Antonio (2001), Kilpatrick et al. (2006)
www.cdc.gov/ncidod/dvbid/westnile/qa/symptoms.htm

Sources for Aedes albopictus:
Infectieziekten Bulletin *16 (2005): 275-76*
Infectieziekten Bulletin *17 (2006): 43-44*
creatures.ifas.ufl.edu/aquatic/asian_tiger.htm
www.flutrackers.com/forum/archive/index.php?t-2607.htm

Aedes albopictus can transmit several viral pathogens.

Lucky bamboo transports can introduce *Aedes albopictus*.

Thirteen misconceptions and half-truths about biological globalisation

1. "Biological globalisation is a natural process; there is nothing new under the sun."
That is only half true. Countless bio-invasions have been and are being facilitated by humans, either intentionally or accidentally. This process is enhanced by climate change and the globalisation of the economy.

2. "Biological globalisation and bio-introductions do more good than harm."
This is true and not true, depending on categories and definitions:

* Globalisation of crops has by and large been beneficial. Across the globe, present-day agriculture is largely based on introductions of highly productive exotics. The economic benefits of these exotics far exceed the costs and the damage involved. Even for nature there are benefits, insofar as high crop yields leave more land for natural habitats. However, most introduced crops remain under human control and are not generally considered "invasive".
* Some introduced tree species, though not invasive, do much environmental harm, such as forest fires and water abstraction.
* Several introduced livestock, fish and shellfish species, such as feral pig and Japanese oyster, are invasive in some regions and do much ecological harm.
* Among non-productive introduced species, some did spread widely, and some of these became genuine invaders, harming nature, the economy and/or public health.

On balance, we cannot say that bio-introductions generally do more good than harm.

3. "Biological globalisation enriches nature."
Half true. By definition, every bio-invasion makes an ambiguous contribution to biodiversity. Positive: it enriches local biodiversity. Negative: it reduces diversity between biotas. However, one invasive species may displace one or more indigenous species, particularly on islands and in lakes and rivers. Bio-invasions rank among the four primary causes of global biodiversity decline.

4. "Bio-invasions only harm the biotas of islands and the relatively isolated Australian continent."
Not true. Lakes and rivers are equally vulnerable, and even on large continents invasions have caused major damage, often permanently. In the US alone, the costs of damage and control are an estimated $120 billion per year.

5. "Biological globalisation is always harmful."
Not true either. Many crop and livestock introductions have been beneficial to agricultural production and the population without causing much damage to nature, although agriculture of course always takes the place of natural habitats.

6. "Nature itself will deal with bio-invasions."
Half true. Invasive exotics often show a boom-and-bust pattern meaning they are "tamed" by natural enemies or resource scarcity, but by then they may already have caused significant damage. Furthermore, taming often does not occur at all, as has been observed on many islands and in many lakes. Even on continents, the invader may remain unchecked. For instance, a fungus from Japan has almost entirely eradicated the American chestnut. Introduced natural enemies can sometimes do the job, but polyphagous species can do more harm than good, and monophagous predators may not be available.

7. "The bulk of bio-invasions are now behind us."
Wishful thinking. The rate of invasions is still accelerating, and huge reservoirs of potential invaders are waiting to hitch a ride to distant locations. Some of these will do harm to biodiversity, public health and/or the economy.

8. "Biological globalisation cannot be halted in this age of globalisation."
Half true. Most invasions driven by climate change and by breakdown of natural barriers cannot be stopped, but many invasions caused by long-distance traffic and transport can be. For instance, following the SARS outbreak in China by the end of 2002, the virus was rapidly contained in a globally coordinated campaign. In those few countries where it had already arrived, it was isolated and then eradicated. Global hygiene is not an illusion.

9. "Aren't there good rules in place to prevent bio-invasions?"
That may be true for agricultural and livestock pests, much less so for other exotics. Only a few nations (such as New Zealand and Australia) have an overall policy against bio-invasions.

10. "Efforts to combat biological globalisation are harmful to global trade."
Not generally true. Firstly, risks are not distributed randomly among trade flows, but are concentrated on specific flows, mainly flows of living materials, produce and

wood and flows between regions having similar climates. Stringent rules and checks are needed here, which may or may not impede trade. Ballast water is another problem, but this can be regulated without significantly affecting trade. Secondly, *not* preventing invasions can do even more harm to trade, since importing countries may close their borders. Thus viewed, invasion control policies are in the best interests of trade.

11. "Measures to prevent and control bio-invasions will harm developing nations."
Not necessarily. First, the damage to developing nations is less than one may think, since the most significant risks are involved in East-West transports rather than South-North transports. Second, potential reductions in trade can be compensated through *additional* trade in *processed* products. The latter are less likely to cause bio-invasions, as well as more profitable to developing countries. The EU could achieve this by ending the so-called "tariff escalation", viz. by lowering tariff barriers for processed products to the same levels as set for fresh products. That will reduce invasions whilst creating better opportunities for processing industries in developing countries.

12. "Biological globalisation is not a serious problem in the Netherlands."
Not true. Over the last decades the Netherlands has been facing:
* major and very costly epizootics of BSE, swine fever, food-and-mouth and bird flu;
* invasions of such agricultural and horticultural pests as the Med fly, Californian thrips and chufa flat-sedge;
* muskrats continuously undermining dikes;
* giant hogweed locally harming children;
* Japanese and giant knotweed swamping on parks and roadsides and replacing indigenous plants;
* roadside elms suffering from a second invasion of a more virulent strain of Dutch elm disease;
* forests still being overgrown by black cherry;
* horse chestnuts colouring brown from July due to the horse-chestnut leaf miner;
* tidal zones overgrown by the Japanese oyster;
* fens dominated by the pumpkinseed sunfish displacing red list frogs and toads.

Other species seem to have started invading, such as *Dermacentor* ticks (harming dogs and humans), the corn rootworm (maize) and bluetongue (cattle and sheep). Imminent threats are bird flu (fowl and possibly humans), West Nile (birds, horses and humans) and dengue (humans). Many others will follow.

13. "As a small nation, the Netherlands plays only a minor role in biological globalisation"
Not so. As a trading nation and major exporter of living materials, including flowers, livestock and ballast water, the Netherlands has a disproportionately large impact and, hence, responsibility.

Lessons learned and actions needed

Lessons learned

Before examining options for action and policy to control invasive species, we first summarise what can be learned from previous chapters:

1 Bio-invasions are not a new phenomenon, but they are now happening more frequently and on a larger scale, due to climate change and accelerating globalisation.

2 Invasions have been reported in all terrestrial and marine natural and semi-natural ecosystems, as far as Antarctica. Invasive species are found in a wide range of taxa, from viruses and bacteria to fungi; from algae and mosses to ferns and flowering plants; and from protozoa, nematodes, molluscs and arthropods to fishes, amphibians, reptiles, birds and mammals.

3 In the Netherlands, invasions are becoming more and more frequent as well. For example, since 1992 alone, at least 30 species of arthropods have become established in the country, 12 of which were introductions.

4 Williamson's tens rule, about 10% of species introductions lead to invasions, and about 10% of invasions (i.e. about 1% of introductions) develop into a pest, is an underestimate for animals and even for plants.

5 The effects of many bio-introductions are more or less neutral.

6 Some introductions have been largely beneficial, in particular deliberate introductions of crops and livestock. In most countries agricultural production is primarily based on non-indigenous plants and animals.

7 Some introductions are harmful to nature and biodiversity, the environment, the economy and/or public health. With the present high frequency of invasions, the percentage of pests among introduced species does not need to be high to cause significant damage.

8 Bio-introductions are fundamentally different from abiotic (physical and chemical) introductions. An introduced toxic chemical may cause damage for years, decades or even centuries, but in most cases it will sooner or later be decomposed by microbes. It

may also spread, but then usually becomes diluted. By contrast, an introduced species may proliferate and spread widely without being diluted, and will often be there to stay – so may the damage. In other words, introductions carry a *small* risk of causing *widespread and permanent* damage. Therefore, more care needs to be taken over the introduction of a species than the introduction of a chemical.

9 The following ecological factors can contribute to the expansion of a species (Pimentel 2005):

- The species carries few or no parasites and encounters few or no parasites and other natural enemies in the new area (e.g. the muskrat).
- The invasive parasite or pathogen associates with a new host (e.g. the HIV and bird flu viruses).
- Predators settle in a 'naive' biocoenosis (e.g. the brown tree snake on Guam).
- Artificial and/or disturbed habitats are created (used by e.g. agricultural weeds).
- The species is exceptionally successful and flexible (e.g. water hyacinth and sweet-potato whitefly).
  Finally, some species produce chemicals that eliminate natural enemies and competitors.

10 Among the threats to global biodiversity, bio-invasions rank fourth or fifth only after habitat destruction, climate change, exploitation and possibly nitrogen (NO_x and ammonia) input. With all these factors in full swing, the planet is entering an era of high biotic decline and instability.

11 The favourite strategy of nature conservationists, the establishment of nature reserves, offers little protection from bio-invasions. Nor does it offer any protection from climate change.

12 In some cases (e.g. influenza viruses) an invasive pest species will become extinct naturally. If it survives though, the damage may decrease over time, partly due to co-evolution of the invasive species and its host or prey species. But we cannot rely on that. The chestnut-blight fungus, for instance, has all but wiped out the American chestnut, yet the fungus is showing no signs of decreased virulence.

13 The greatest risks occur in the following situations:
- Small and remote regions such as lakes and oceanic islands with endemic species.

- Disturbed ecosystems rich in resources or with strongly fluctuating resources.
- Transportation between matching climate zones. For the Netherlands, these climate zones are North America and East Asia in particular, as well as tropical and subtropical zones in the case of heated greenhouses, buildings and even the streets of cities.

14 If there is no change in policy, the frequency of invasions will grow in proportion with trade flows. Given the increasing trade with China, this part of the world will probably be a source of many invasions threatening Europe and other regions.

15 Costs to the economy occur in a range of sectors, such as agriculture, fishery, forestry, shipping, water management, energy supply and not least public health. The total cost (including control costs) is estimated to be $335 billion per year in the US, UK, India, South Africa, Brazil and New Zealand alone. Of that figure, $229 billion consists of economic costs to agriculture and forestry and the remaining $106 billion is for costs to other sectors, including human and livestock health. This is equal to 5% of the annual GDP in those six countries. If we extrapolate that to a global scale, the damage would be $1,400 billion per year.

16 The economic costs from invasive species in the Netherlands are poorly known. A first estimate, based on the sum of all the known costs of prevention, control, losses and damage, supplemented with estimates of some other items, amounts to €1.1-3.1 billion per year.

17 Among the invaders, some taxonomic groups generate much more costs than do other groups. In the US, mammals (mainly rats and cats) and microbes dominate the scene. But using a somewhat more realistic calculation, the microbes are clearly on top. The lower vertebrates (fishes, amphibians and reptiles) generate much lower costs. Among the invertebrates the only group next to the arthropods that causes substantial damage is the molluscs (mainly zebra mussel and Asian clam) and probably the nematodes.

18 In the Netherlands by far the most costly invaders are the viruses affecting human and livestock health, accounting for more than 50% of the total costs from invaders. They are followed by the wide range of organisms attacking crops. Less costly invaders are the muskrat and exotic cockroaches.

19 Much of the economic damage, for example by (invasive) agricultural pests and the muskrat, is not systematically registered in the Netherlands.

20 In the Netherlands, the percentage of harmful species among invaders from distant regions (particularly North America and East Asia) is much higher than the percentage of harmful species among invaders from European countries. This is presumably because species from distant regions are accompanied by fewer natural enemies than species from Europe.

21 Damage to environmental quality is primarily caused by:
- preventive decontamination of live export and (rarely) import material;
- chemical methods to eradicate or control invaded species.

22 The damage to nature and biodiversity consists mostly of the displacement of indigenous species through predation, parasitism and competition. In some cases ecosystems are disrupted through overgrazing, erosion, forest fires, drought and nitrogen fixation. Displacement of species to the level of extinction is a frequent event on islands, in lakes and in rivers, but is thus far a rare event on continents and in oceans (leaving aside the exterminations by prehistoric and historic *Homo sapiens*). This is also true for the Netherlands, where no species have completely disappeared as a result of bio-invasions, although this has been known to happen locally. The pumpkinseed sunfish, for instance, is causing local extinctions of rare frogs and toads. Ecosystem damage has been small, mainly due to intense efforts to combat the muskrat, but the Japanese oyster is changing estuarine ecosystems. The efforts being made to control the pumpkinseed sunfish and muskrat are also causing damage to the indigenous fauna.

23 Bio-invasions will take place more and more frequently in the future, in part due to climate change, and in part to the expected increase in volume of transported material and the speed of travel. The world will face many thousands of invasions, creating millions of new ecological interactions, few of which are predictable. Problems will continue to mount even if all invasions were to cease, because many bio-invasions have a lag phase that lasts for decades. A fair proportion of introduced species are 'dormant invaders'. In addition, some of the species already considered pests will continue to expand.

24 As the number of bio-invasions increases, so will the damage. Some examples:
- The US livestock industry has over 60 potentially invasive pathogens and parasites to worry about.
- In New Zealand, 47% of all plant species are non-indigenous, and invasions from many more aquatic and terrestrial plants and animals are feared (Green 2000).
- Of the 18,000 species reported by the IUCN as being

Sea-ships as well as their cargo and ballast water are the main vehicle for aquatic bio-invasions.

under global threat, a third are partially or primarily threatened by bio-invasions.

- Bio-invasions are considered the fourth or fifth most important threat to global biodiversity.

25 In the Netherlands, developments are moving at an alarmingly rapid rate. On average, new species becomes established at the following rates:

- one freshwater species every year;
- one saltwater species every year;
- one plant species every six months;
- at least one arthropod species every four months.
  In the Rhine fauna, a stealthy revolution has taken place over the last two decades.

26 The following groups of *terrestrial* exotic species appear to be most worrisome for the Netherlands.

- The global threat of new emerging diseases, mainly zoonoses, the greatest risks being posed by virulent viruses transmitted from person to person by sneezing and coughing. Such a virus might develop from the H5N1 bird flu virus that has become endemic in China.
- Other pathogens in plants, livestock and humans, as well as their vectors. Particular threats are West Nile disease (vectors present, virus possibly present in migratory birds), *Aedes albopictus* (recently found locally, though not yet with the dengue virus) and the sheep disease bluetongue (vectors present, virus recently found). Other potential threats are the

hanta virus (originally from Asia), transmitted by the bank vole, and arthropods harmful to agriculture (such as *Thrips palmi*, new potato diseases and pests, as well as new maize diseases and pests like the Western corn rootworm), greenhouse horticulture and ornamental plants.

- Plants and their associated pathogens and parasites introduced by garden centres, that might spread and harm nature and biodiversity, agriculture, forestry, gardens or waterways.
- Beetles affecting the wooden parts of houses and other buildings.
- New, more harmful varieties of pests and pathogens already established, e.g. multi-drug resistant varieties of the tuberculosis bacterium and pesticide-resistant varieties of the Colorado potato beetle.[94]
- Hitching a ride is relatively easy for small organisms (including small reproductive stages like eggs, seeds and spores).

27 The following groups of *aquatic* exotic species appear to be most worrisome for the Netherlands.

- toxic algae;
- species introduced for aquaculture, as well as their pathogens and parasites;
- Ponto-Caspian species, particularly small crustaceans displacing the indigenous fauna;
- seaweeds, fishes and crustaceans escaping from fish farms;

Airplanes can spread invasive species, including pathogens, across the globe within a matter of days.

- species released into the wild by aquarium owners;
- species hitching a ride with plants in water, such as *Aedes albopictus* (see above);

28 However, the most dangerous species may be a well known but dormant species, an unknown species or even a species that does not yet exist. Before 1980, nobody had heard of HIV/AIDS, SARS and BSE, nor had anybody expected the global expansion of the *Caulerpa* seaweed. And who, except for a few experts, had foreseen the devastating invasion of the ctenophore *Mnemiopsis* in the Black Sea and the zebra mussel in North America?

29 Nowadays, most bio-invasions are a side effect of 1) climate change, 2) globalisation of traffic and transport, and 3) the breakdown of natural barriers. These three categories require entirely different approaches.

30 Invasions related to *climate change* cannot be halted unless the species needs to cross a habitat barrier. In most cases we should not even try to halt them, since this would trap the species in an unsuitable climate, endangering its survival, at the same time creating incomplete and unstable biotas that are poorly matched to the climate. Furthermore, the risk that these invaders will dominate permanently is limited, since sooner or later they will be joined by natural enemies checking their numbers. In those cases where species cannot adapt their range to the changed climate because of a distributional barrier, it may even be wise to help these species by translo-cating some of them, or by creating ecological corridors for all of them. But of course, this should not be done across ancient barriers such as oceans.

31 Most invasions resulting from the *breakdown of a natural barrier* (e.g. the construction of a road through a continuous forest belt, or the construction of a canal connecting separate bodies of water) cannot be prevented either. Fortunately, most of these invaders, too, will be joined by their natural enemies, reducing their ability to gain dominance for a long period of time. Again, in most cases the best strategy may be to allow 'natural' succession to run its course.

32 By contrast, species invading with *trade and traffic* will often *not* be joined by many natural enemies and will therefore have a greater opportunity to develop into pests. Fortunately, these species are less difficult to halt since:

- they often have to cross natural barriers and bottle-necks such as seaports and airports;
- many past harmful invasions resulted from deliberate introductions or even "pushing", and such mistakes can be avoided in future;
- even 'flash' invasions can sometimes be controlled. For instance in 2003 a globally coordinated campaign against the new emerging disease SARS contained and eradicated the virus within a month;
- risks are often concentrated in specific flows, particularly the flow of living materials between countries with matching climate zones.

Countries like New Zealand and Australia, having learned tough lessons from previous invasions, have been successful in reducing the number of invasions.

33 The assumption that those species invading in relation to climate change or the breakdown of a natural barrier will be joined by their natural enemies does not hold for viruses. Viruses have few natural enemies, their main enemies being the antibodies and leucocytes of animals, which will not join them when they travel. Since climate related invasions are hard to stop and control, the best strategy may be improving hygiene and enhancing immunity of animals and humans.

34 Generally speaking, there are few natural defence mechanisms against bio-invasions. One mechanism is the highly developed immune system in animals, which has an innate (broad) and an acquired (specific) subsystem. Both subsystems can, in principle, be activated. It is important to create strategic stockpiles of vaccines against dangerous exotic pathogens that cause disease in humans and animals. However, vaccines are not available for all pathogens, and wild animals are generally not vaccinated.

35 As for ecosystems, a potentially effective method is an organized counter-invasion by a pathogen, parasite or predator, but these are not always available and introducing them can sometimes do more harm to non-target species than to the target species. Even monophagous pathogens or parasites sometimes do more harm than good.

36 Of all the human influences on the environment, bio-invasions are among the least predictable. The ecological theory is still barely able to predict which species will be invasive and what the effects of invasions will be. Some species do not become invasive until decades after they were introduced; others – like the zebra mussel – quickly reveal themselves to be harmful species, but by then have spread so widely that they can no longer easily be controlled. The example of the Japanese oyster in the Netherlands shows us that predictions by experts can turn out to be erroneous. Although theoretical progress has been made in this field (see for example Ricciardi & Rasmussen 1998) significant uncertainties remain.

37 There are four arguments for a restrictive policy based on prevention and the Precautionary Principle:

- The damage from introduction may be large and widespread.
- The damage may be irreversible. In many cases it is not possible to predict the damage.
- Even if eradication or control is possible, prevention is often more cost effective and less harmful to the environment.

Therefore, species should not be introduced unless the effects are reasonably predictable and considered acceptable (see also Shine *et al.* 2000).

Such a 'no, unless' approach is already commonplace

The huge Dutch flower auctions, although well surveyed, are a global risk factor for invasions of plant pests and pathogens.

in legislation on the approval of new chemical substances and genetically modified organisms.

38 The Netherlands is not a minor player in biological globalisation. It produces and exports large quantities of live material, including livestock, vegetables, fruit, potted plants, cut flowers, flower bulbs and living propagation material, mainly plant seeds, seed potatoes, sperm (especially from bulls), broilers and piglets. The country is even the world's biggest exporter of broilers. The Netherlands also has giant international flower auctions where cross-contamination can occur. In addition, the country exports huge quantities of ballast water. The Netherlands is therefore not only a susceptible object but also among the major drivers of bio-globalisation.

39 Looking ahead, with Rotterdam as the world's largest seaport outside Asia, and China having eight large and fast-growing seaports, China–Rotterdam may well become the world's leading two-way avenue for bio-invasions.

40 The main risk factors of bio-invasions are:[95]

- Transportation between *similar climate zones*, crossing classical distribution barriers. Usually this involves transport by ship or plane, rather than by car. Transport between Europe and North America or Japan is more risky than between the Europe and the tropics.[96] Likewise, inter-tropical transport is more risky than transport between these tropical and temperate zones.

- Transport of *living material:* in animals, major risk factors are the shellfish trade and the aquarium trade; in plants, water or soil associated with plants (potted plants, tree cultivation products, seed potatoes) present a more significant risk factor than the above-ground parts (e.g. cut flowers).

- Transport of *agriculture and forestry products* that have undergone little processing, such as fresh, cooled or frozen fruit, meat, dairy products, wool, tobacco and wood, including wooden packages and pallets and ballast wood.

- Transport of *ballast water.*

- Transport as *fouling of ship's hulls.*

- Locations where flows of living materials intersect, such as *harbours, airports, livestock markets, fruit & vegetable and flower auctions and garden centres.* Major points of entry are disturbed ecosystems rich in resources.

41 In the Netherlands, the most important vehicles appear to be ballast water, ship's hulls, flowers, plant material, livestock and propagation material (sperm, seeds, seed potatoes).

42 Examining the present policies, we observe four significant inequalities:

- Although bio-invasions rank fourth or fifth as threats to global biodiversity, they still receive little attention in Europe and the Netherlands, even among NGOs.

- The introduction of new chemical substances and genetically modified organisms is not a greater risk than the introduction of exotic species. On the contrary, the effects of exotics are often less predictable, less reversible and less controllable than the effects of chemicals. Yet bio-introductions are *less* strictly regulated in most countries.

- Those regulations that exist focus on introductions of agricultural, forestry and fishery pests, as well as human and livestock diseases. Very few regulations are in place for introductions that may harm biodiversity (Shine *et al.* 2000).[97]

Today, these inequalities can hardly be justified.

43 Another inequality can be found in the relevant research efforts. Most research still focuses on the ecological effects of bio-invasions; a much smaller proportion is focused on distribution mechanisms, prevention and economic effects (Bossenbroek *et al.* 2005). In addition, there is still little multidisciplinary research involving experts in ecology, economy, medicine and law.

44 It should not go unmentioned that many exotic pathogens and parasites of humans, livestock and crops are potential weapons of mass destruction that could fall into the hands of bio-terrorists.

45 All things considered, *laissez-faire* is a risky and irresponsible strategy.

Actions needed

Based on these lessons, we formulate the following recommendations:

1 Most climate-driven invasions cannot, and should not, be halted. Where such invasions are stopped by habitat barriers (not being an ancient barrier such as an ocean), they should even be helped by translocating some species or creating ecological corridors for all species. Most invasions resulting from the breakdown of a natural barrier cannot, and should not, be halted either.

2 Many invasions related to travel and transport can, and should, be halted. The most sensible policy appears to be the one contained in article 8 (h) of the 1992 Convention on Biological Diversity: 'to prevent the introduction of, control or eradicate those alien species which threaten ecosystems, habitats or species', though only 'as far as possible and appropriate'. Many parties have adopted this view. More specif-

Garden centres are a key source of plant invasions and need to be regulated.

ically, there are three hierarchical guiding principles:

- Prevention of invasions.
- Detection and, where suitable, rapid eradication efforts to prevent settlement.
- For species yet established: containment or control.

3 Prevention is best focused on the priorities given in item 40 above and the pathways listed in item 45 above.

4 In the importing country, prevention measures should include not only border controls, but also optimal use of the 'natural resistance' of biotas: limiting habitats that are disrupted and resource-rich, and perhaps even activating indigenous predators, parasites and pathogens.

5 The opportunities for eradication are best in the early stages of the introduction. The longer one waits, the more difficult and expensive eradication becomes. 'Nipping in the bud' is the best strategy. It is unwise to wait for the species to cause harm or to gamble that it will not do so.

6 A 'no, unless' policy is also advisable for intentional introductions.[98] However, as more solid predictive theories and methods become available, we can rely more often on risk analyses. Even then, however, the Precautionary Principle remains of key importance.[99]

7 Control must be specific, targeted at the invasive species only. In principle, the best way is to introduce monophagous natural enemies, if they exist. Even then, however, careful testing beforehand is essen-

tial, especially on efficacy.

8 Worldwide, the challenge is to decouple economic and social globalisation from bio-globalisation, and, more generally, to replace free trade with safe trade. Basically, decoupling requires different treatment of living and non-living material, including meat, wooden pallets etc. The trade in living organisms, fresh products and wood must be regulated more tightly. High-risk trade in living material (particularly trade between distant countries with similar climate zones) will need to be more strictly regulated and controlled. In some exceptional cases, reduction of the specific trade flow us *such* will have to be considered. This already happens occasionally in the case of agriculture and livestock pathogens and pests.

9 Preventive measures are not without socio-economic problems. For instance, sanitary import barriers may be detrimental to exporting countries. However, this applies less to developing countries because most have a tropical or subtropical climate. For countries with a temperate climate, imports from tropical and subtropical countries are less risky than imports from other countries with a temperate climate. Furthermore, industrialised countries can compensate developing countries with additional trading opportunities for processed products. This works both ways: for both 'North' and 'South' there is a reduced risk of bio-invasions and for 'South'

In the battle of bio-invasions, ballast water must be a priority target.

there are more opportunities to develop processing industries. Where trade *among* developing countries is expected to grow, even more stringent measures may be needed in their mutual interest. However, appropriate resources are often lacking.

10 Another problem of prevention (although this applies equally to health care, for example) is that politicians get less political credit for prevention than for control. Likewise, most companies make less profit from prevention than from control. Politicians and private parties involved in trade must be made aware of their responsibility for the long-term costs and risks. One way to raise awareness in importing and exporting companies may be to make them legally liable for the damage from species introductions and to apply the Polluter Pays Principle. A preventive effect may be expected.

11 Just like prevention, eradication of established exotic species can cause problems. In some cases it is not technically feasible, it is usually costly, and it can lead to protests from the general public. Local residents

may have grown attached to an exotic tree species; some conservationists believe exotic species have equal rights to be protected; some critics consider the fear of exotic species to be a form of xenophobia; and hunters may be unwilling to part with game. All the more reason to shift the emphasis toward prevention, precaution and 'nipping in the bud'. But it is also a reason to carefully weigh up the pros, cons and risks of those exotic species already established as well as to hold a broad debate on the issue.

12 Based on this book, we formulate the following *policy agenda*:

a End the unequal response to the risks of bio-invasions compared to the risks of habitat destruction, climate change, chemical substances and genetically modified races. There is some catching up to do here.

b Raise awareness of the risks of bio-invasions through public communication campaigns targeted both at the general public and specific groups. Such campaigns should spread the message that releasing

Biological globalisation

exotic species into the wild is just as risky as (or even more risky than) dumping chemical waste. Greater attention should be given to the paradox of scale: bio-invasions can lead to a decline in biodiversity on a global scale, but can lead to increase in biodiversity on a local scale. Priority target groups are customs officers, tourists, garden centres and the aquarium trade.

c Formulate, in consultation with stakeholders, codes of conduct for all the key players. More attention should be given to all flows of living materials, as well as wood and wooden packing materials, across national borders. Loopholes in legislation must be closed, particularly where these concern the aquarium trade, garden centres, the trade in wood, wooden packing materials and ballast water.[100]

d More stringent policing of existing regulations, focusing on high-risk flows, particularly those to and from areas with matching climates. For Northwestern Europe these include the US, Canada, China and Japan.

e Examine possibilities for applying the Polluter Pays Principle. This could be done by tariff differentiation, imposing higher tariffs on high-risk flows; or by making importers or exporters liable for possible damage from bio-invasions. We can expect such measures to have a preventive effect. However, they may require changes in the General Agreement on Tariffs and Trade (GATT) (Perrings *et al.* 2005).

f Stimulate the reporting of bio-invasions by introducing a duty to report, where necessary coupled with provisions to compensate those people who may suffer damage from reporting. In the Netherlands, such regulations are already in place for contagious animal diseases, but not yet for contagious plant diseases. Compensation is also vital to the success of eradication campaigns.

g Reduce the number and areas of disturbed, resource-rich and resource-fluctuating habitats where exotic species often settle and start spreading.

h Provide less legal protection for exotic species of plants and animals.

i Where possible, give indigenous species priority when planting trees and shrubs.

j Actively participate in developing multilateral policies within the framework of the Convention on Biodiversity CBD, the Convention on International Trade in Endangered Species of Wild Flora and Fauna CITES, the UN Food and Agriculture Organisation FAO, the World Trade Organisation WTO, the International Maritime Organisation IMO, the World Health organisation WHO, the World Organisation for Animal Health OIE, the International Plant Protection Convention IPPC, and the Ramsar Convention on Wetlands. The focus should be on prevention, eradication and control of invasive species.

k Test developmental aid policies for any intentional or unintentional promotion of bio-introductions.[101]

l Provide bilateral aid to developing countries with high levels of biodiversity and endemism, including island states, in order to prevent bio-invasions and reverse them where necessary and possible. Aid should include finance as well as capacity building.

m Anticipate increased trade between developing countries.

n Support the use of indigenous species in agriculture, forestry and fishery in developing countries. Thousands of tropical tree species, for example, have never been tested for their usefulness in forestry (Bright 1998).

13 Institutional changes are also required in many countries, including the Netherlands, for the prevention of bio-invasions:

a Many specialised agencies are involved in bio-introductions, each focusing on a single issue: human pathogens; veterinary pathogens and parasites; plant pathogens and parasites; and endangered species. Such a segregated approach is inefficient and ignores the countless linkages between all these biological flows. An integrated approach is needed, based on a 'One Nature' concept (Karesh & Cook 2005).

b A first step could be the creation of a single coordinating agency for the safety and bio-safety of border traffic. Further efficiency gains can be achieved by combining this with border checks for drugs, terrorism and arms.[102] Instead of searching for needles in the haystacks of all different flows, it is much more efficient to identify high-risk flows and focus on them.

c It is also worth considering the creation of a Bio-security Council that would advise on *all* introductions of new species and varieties, whether by import, conventional breeding or genetic engineering.[103] In the US a first step was made in 1999 with the creation of the National Invasive Species Council.[104] Another option would be a body with enforcing power, in which case a Bio-security Authority would be appropriate.

14 We conclude with the following research agenda:

a Stimulate *multidisciplinary research* on the analysis, prevention and control of bio-invasions, involving ecologists, experts on plant and animal diseases, medical epidemiologists and economists. Zoonoses such as bird flu have shown that medical epidemiologists cannot avoid cooperating with ecologists, ornithologists, veterinary experts and economists. The time for mono-disciplinary research on bio-invasions is over (see also Bryan 1999, Karesh & Cook 2005).

b   Sustain the necessary *specialist expertise* on the taxonomy, detection and control of all potentially invasive taxa. This requires greater cooperation at the European level. In the Netherlands some of the knowledge existing in fields such as rare plant diseases, has been lost as a result of budget cutbacks.

c   Improve the *monitoring of invasive species*, particularly at seaports and airports. The goal is to provide *early warning*. When a suspect species is found, it must be quickly eradicated if possible, unless there are good reasons not to do so. If eradication is impossible or undesirable, the species should be adequately monitored, including estimates of possible effects and damage. Such a study may need to continue for years, with the aim of gaining more insight into bio-invasions and their effects.

d   Create national *databases and websites* on invasive species, following the examples of the US, UK and Australia.

e   Develop improved *detection techniques* for small invasive species, to be used at border checkpoints. Also use high-tech physical, chemical and biological techniques (including DNA techniques) and equipment, trained dogs and pigs, or even parasitoid wasps.[105]

f   Initiate more research into ecosystem *susceptibility* to bio-invasions.

g   Initiate research into options to reinforce *biotic resistance* against invasions by activating and facilitating native (!) pathogens, parasites and predators.

h   Intensify research into *monophagous natural enemies* of invasive species.

i   *Organise better monitoring and analysis of the costs incurred* to various economic sectors. The gaps in knowledge are still considerable.

j   Make detailed analyses of the invasion process and its costs by taxon, geographical origin and pathway of introduction.

k   Further develop *risk analysis methods and procedures.* This includes risk assessment, risk management and risk communication. Assessing risks means looking at the size and nature of the potential adverse effects of a proposed introduction as well as the likelihood of it happening. It should identify effective means to reduce the risks and examine alternatives to the proposed introduction (Shine 2000).

l   Encourage *international consensus* on this subject; the OECD has already taken the first steps towards achieving such a consensus on the calculation of costs from invasions.

m   Apply such *risk analyses* to a wide range of species associated and to specific flows of trade: which species are likely to be introduced into the country, from which regions and along which pathways? What are the economic, ecological and social risks of this? Which flows cause the greatest risk, what are the critical control points[106] per flow, and where and how can prevention and control of bio-invasions be executed most cost-effectively? What are the chances of those risks? Policymakers must be offered clear options, including the costs, benefits and risks avoided for each option.

n   Develop *methods and techniques* for importing and exporting companies and their customers to *reduce the risks of introductions*[107] and, based on such methods, to develop *systems of best practise, transparency and reporting*. Ideally, importers and exporters will consider prevention of bio-invasions to be a full-fledged element of socially responsible entrepreneurship. This is not an illusion. Recently, Dutch gardening centres voluntarily decided to stop the trade of water pennywort on request of the waterboards.

The Netherlands – which calls itself a 'distribution country' – is not a minor player in world trade, especially not in living matter and ballast water. This involves particular risks and responsibilities, as well as opportunities. But obviously, national efforts alone can have only limited effects in the battle on biological invasions. An effective policy needs to be European and global.

CHAPTER 1.
Introduction

[1] The term "biological globalisation" was introduced - as far as we know independently of one another - by staff mem-bers of the University of Texas at Austin in 1995 (http://www.eco.utexas.edu/~archive/chia-pas95/2001.09/msg00170.html); by Andersen (1995); by Bright (1998: 200); by McNeill (1999); and by Van der Weijden (2003). McNeill reserves the term for the post-Columbian transoce-anic exchanges. Related terms are "biotic homogenisation" (McKinney & Lockwood 1999), "biotic globalisation" (Davis 2003) and "decompartmen-tation of the world" (Bol & De Hollander 2001). The term "biological globalisation" is slowly invading the planet. An internet-search on January 16th, 2007, yielded 1731 hits. However, the brief term "bio-globalisation" may cause confusion since it is sometimes used to refer to global genetic engineering.

[2] Perhaps the "spontaneous" invasions of the fulmar and the collared dove were facilitated by changes in methods of fishing and agriculture, respectively.

[3] Pliny also reports the lynx and bear as being used in the circus, but these species came from Europe, possi-bly even from Italy itself. Remarkably, he reports that the import of animals from Africa for reasons other than circus games was banned; presumably this was because of the risks. In fact the species had been selected for being dangerous to humans.

[4] Previously, in 1927, Elton had laid the foundations for the ecology of animals with his *Animal Ecology*.

[5] One book takes a sceptical approach to bio-invasions (Theodoropoulos 2003), describing bio-invasions as an enrichment of nature. However, this book is not taken very seriously in the specialist literature (Dahler 2004, Tye 2005).

[6] The introduction of infectious diseases from elsewhere is, in fact, one of the oldest fields of study in medicine. In his *Al-Judari wa al-Hasabah*, the Persian physician Rhazes (AD 864-930) deftly describes the epidemiolo-gy and transmission routes of smallpox and even sug-gested transmission mechanisms (Bol 1999b). After the start of the plague epidemic in 1346, the Arab scholar Ibn-al-Wardi was quick to describe the trans-mission history of the disease (McNeill 1976). And in 1352 the Italian author Boccaccio, in his introduction to Decameron, gave a poignant description of the plague, speculating on the means of transmission. In the early 16th century another Italian, Fracastorio, described the syphilis epidemic in Europe, postulating that infectious particles were the cause of the disease. British physician John Snow in 1854 gave an accurate description of the epidemiology of cholera, some 25 years after the disease had made its appearance in Western Europe.

CHAPTER 2.
The nature and scope of bioglobalisation

[7] Sometimes barriers are less effective than they appear to be. For instance, the tropical Atlantic appears to be an effective barrier to species from cold waters. But *Calanus finmarchicus*, the most abundant copepod crustacean in the plankton of North Atlantic seas, can get to the Indian Ocean near Madagascar because the cold northern currents dive downward and travel below the warmer surface waters of the tropical seas (Elton 1958).

[8] According to Wallace (1876: 188), the line continued on between Borneo and Celebes, but he added that Celebes is, zoogeographically speaking, more of a tran-sitional area and also has unique species such as the babirusa, a highly divergent type of pig. The current practice is to place the islands east of Wallace's line in a separate biogeographical region known as Wallacca. This region is considered one of the world's 25 hotspots for high but endangered biodiversity.

[9] This brief survey has partly been derived from Reumer (2005).

[10] The mere five animal species found in AD 1917 on Easter Island were: a lacewing fly, a fly, a beetle, a water beetle and a land snail. Presumably, though, the flora and fauna had been much less poor before AD 1200, when Polynesians colonized the island and started hunting and cutting forests.

[11] A few birds on continents, including the ostrich, have also lost their flight capacity. Apparently they are strong and swift enough to cope with predators.

[12] Williamson (1997) calculated the "shrinkage" of the body size of mammoths on Jersey during the last ice age to be 1% per century. In this light, it comes as no

surprise that we have so far seen only minor changes in introduced species.

[13] These figures should be used with great care, because different authors give different figures for the same region. Heywood (1989) for instance gives a figure of 1200 to 1300 native species for Hawaii while Pimentel *et al.* (1999) arrive a total of 2690 native species. For exotic species the figures are 228 for Heywood versus 946 for Pimentel *et al.* Definitions can make a major difference. Vermeij (1996) states that figures from diverse biotopes and regions should not be compared with one another.

[14] Source: http://earthtrends.wri.org/text/biodiversity-protected/feature-18.html

[15] Pimentel *et al.* (2002) have provided figures for invasive plant, animal and microbe species in six different countries. They estimate the total number of exotic species to be 120,000. The percentage of exotic species varies greatly: from 1% in Brazil, 3% in Australia, 6% in the US and 7% in South Africa, to 19% in India and 53% in the UK. However, these figures cannot be taken seriously. For instance, a figure of 27,515 plant species (including 26,000 exotic species) is given for the UK, but this number is much too high even if we include plants which are imported but not introduced (Table 2). The percentage of exotic species in Brazil appears to be mainly based on the figure of 11,605 exotic plants divided by the total of 55,000 plants plus 1 million arthropods! And this is not the only mistake in this pioneer work, with excellent chapters alternating with weaker ones. For instance, in the chapter on invasive human disease, high infection rates (e.g. for HIV) are given for entire populations based on figures for distributions in subpopulations.

CHAPTER 3.
History, drivers, mechanisms and pathways of bio-invasions

[16] *Vinland* was described in a manuscript compiled about AD 1387 called *Flateyjarbók*, or *Flat Island Book*.

[17] The northern regions of the Old and New World with cold and temperate climates have so many species in common that some biogeographers combine the Nearctic (North America) and the Palaearctic (Eurasia minus tropical Asia plus Northern Africa) into one super-bioregion, the Holarctic.

[18] 135 Years ago a trip around the world was calculated by Jules Verne to take 80 days, which at the time was considered incredibly fast. A trip to the opposite side of the globe would then take 40 days. Today such a trip takes only one day, which is 40 times faster.

[19] The species exchange between the Red Sea and the Mediterranean is called the Lessepsian Interchange, after Ferdinand de Lesseps, the engineer of the canal. Some of the species involved are fish that simply swim from one sea to the other while others are sea floor dwellers that attach themselves to ship hulls or fish-nets. When these nets are cleaned, the species end up in the harbour and form propagules for invasions.

[20] These four exceptions to the tens rule are: edible crops in Great Britain; birds in Hawaii; insects used for biological pest control; and mammals on continental islands. Williamson provided plausible explanations for all four of these cases. The tens rule allows us to use the metaphor of the iceberg, only 10% of which is visible above water. For example invasive species are the top of the iceberg of all introduced species.

CHAPTER 4.
Patterns of bio-invasions

[21] Pimentel *et al.* (2002) give a higher figure for the US, the UK, Australia, India, South Africa and Brazil together. They estimate the total number of exotic plants, animals and microbes to be 120,000. Of those, 20%-30% would be (serious or non-serious) pest species. But again, their approach is too crude to be taken very seriously.

[22] The introduction of the starling was part of a project to populate Central Park with every bird species mentioned by Shakespeare.

[23] A conspicuous example of a high intrinsic rate of natural increase among invasive birds is the collared dove *Streptopelia decaocto*. This bird lays up to five clutches of eggs per year, and offspring from early clutches sometimes begin to breed the very same year (SOVON 2002).

[24] This is how black cherry could develop into a forest pest in the Netherlands, where a parasitic fungus was absent; furthermore, black cherry benefits from a symbiotic fungus that does occur in the Netherlands (Van der Putten & Rienks 2004).

[25] Whereas the clover in New Zealand had exploded following the introduction of the honeybee, it was driven back since 1996 by another invader, the clover root weevil *Sitona lepidus*, native to Europe and North Africa.

[26] Remarkably, the Nile perch is much less dominant in Lake Albert – where it was introduced in 1955 – than in Lake Victoria.

[27] The evolutionary mechanisms are: selection, cross-breeding (recombination) and mutation. Selection and crossbreeding can cause rapid changes, while mutation is a slower process. Because invasive populations have a relatively small genetic variation, the rate at

which genetic change occurs may drop; this is known as "genostasis". Such genetic contraction has been shown in invasive insects, bird, snails and plants (Williamson 1997).

28 Elton (1958) mentions the example of an asparagus fungus that does not cause epidemics in European asparagus crops; yet, after invading America, it threatened the entire asparagus crop there. The damage in America gradually lessened, however – not through pest control but through genetic adaptation of the fungus, genetic adaptation of the asparagus, or both. Commercial growers, though, rarely have the patience to wait for such adaptation.

29 MacIsaac *et al.* (2004) present a model to predict the distribution of species over lakes; the model considers the extent to which species attach themselves to fish-nets and follow the routes of fishing vessels.

30 http://issg.appfa.auckland.ac.nz/database/welcome/prediction.asp

31 Williamson (1997) noted that there is no consistent relation between body size and the rate of reproduction and that species with a low reproduction rate can also be successful invaders. Examples are the northern fulmar and *Homo sapiens*. However, he does not discuss the relation between body (propagule) size and invasion success as such. This hypothetical relation still has to be tested empirically.

32 Darwin was also amazed to find that humans and animals can spread disease without falling ill themselves.

33 An example of invasion from one island to another is the introduction of the soil-dwelling New-Zealand flatworm *Arthurdendyus triangulatus* in the British Isles in the 1960s. The species probably hitched a ride with a plant (and accompanying soil) sent from New Zealand to a British botanical garden. The flatworm predates on earthworms, and can even wipe them out completely, thereby damaging ecosystems (including agro-ecosystems). The species has settled in Scotland and Northern Ireland and may also gain a foothold on the European mainland (Bright 1998, DEFRA 2003). If it settled in the Netherlands, it would not only damage the soil fertility there, but also the population size of meadow-breeding birds such as the globally threatened black-tailed godwit *Limosa limosa*, which is highly dependent on earthworms during the breeding season.

34 The virus that causes Black Sigatoka disease in bananas originates in the Fiji islands, although it may have arrived there from a continent. See: http://www.hort.purdue.edu/newcrop/morton/banana.html#Diseases. The virus has the potential to become a global threat. Genetic variation in bananas is negligible because the most commonly grown cultivar, Cavendish, reproduces asexually, so resistance cannot be introduced through crossbreeding (at least not using classical methods). At this time the future of global banana cultivation remains uncertain (Van Hengstum 2003).

35 The patterns mentioned must be handled with care; according to Vermeij (1996) there is as yet little supporting empirical evidence.

36 Jeschke & Strayer (2005), working on vertebrate invasions, rejected the concept of European ecological imperialism because they do not find any consistent difference between the success rate for introductions from Europe to North America and the success rate for introductions in the opposite direction. The concept does apply for mammals, but for fish it is the opposite. Introductions from Europe were dominant until about AD1900, but since then more introductions have occurred in the opposite direction. At the moment the score is 70-70. So the concept has historical rather than biological and predictive value.

37 The threat of erosion loomed because the native vegetation was often unable to resist grazing by the introduced livestock. In this regard, it may have been a good thing that grazing-tolerant species were introduced from Europe, as Crosby also states.

38 The rule that high densities coincide with high infection pressure, does not necessarily apply to exotic species, at least not in the early phases of the invasion. The main reason is that quite a number of these species show swamping behaviour is that they got rid of many of their natural enemies.

39 Major historical phases were the rise of animal husbandry in the Mesolithic (stimulating the development of zoonoses); the development of towns in the Neolithic (increasing the infection pressure); and the development of a network of cities during the Roman era (facilitating rapid and large-scale spread of pathogens causing pandemics). Similar developments occurred in South and Southeast Asia.

40 Vermeij's warning (1996) that little has been proven so far also applies here.

41 According to MacArthur & Wilson (1967), the small number of species on *small* islands is caused by the relatively high chance of extinction; the small number of species on *distant* islands is caused by the relatively low chance of immigration.

42 Wallace noted that these invasive pioneer species often disappear again after a few years, though they may have formed and spread their seeds in the meantime. He draws a parallel to what happens along rivers and the edges of snow masses and glaciers, where areas of land can suddenly be left barren. This provides temporary steppingstones for plants. He postulates

that plants from the far north have used mountain ridges as steppingstones to spread to the far south.

[43] To the Caribbean Sea, the much colder Atlantic Ocean served only as a limited species reservoir. In the warmer Indo-Pacific area, the number of species per taxon is 6 to 10 times higher than it is in the Caribbean; hence the number of symbioses that were able to develop was much higher.

[44] Alternatives such as biological or electrical anti-fouling methods are being actively researched. A few marine species, for instance *Hydracytinia echinata*, a common hydrozoan, are not grown over by other species. Not unlikely, the substances they secrete may have some use as an environmentally friendly means of combating ship fouling. Fouling organisms are found in virtually every aquatic taxonomic group, including algae, sponges, sea anemones, hydroids, crustaceans, molluscs, bryozoans and tunicates.

[45] Every year, nearly 400,000 muskrats are trapped in the Netherlands, at a high cost. The muskrat nevertheless causes a lot of damage to dams and dikes. Trapping also has adverse side effects: some 15,000 non-target animals are caught accidentally, including highly valued species such as the crayfish *Orconectus limosus*, polecat *Mustela putorius*, little grebe *Tachybaptus ruficollis* and water rail *Rallus aquaticus* (LCCM 2004).

[46] In 16th century Europeans generally believed that cures for diseases were to be found in the same region as where the disease came from. For instance, drugs for treating syphilis were sought and found in the West Indies: gaiac and sarsaparilla. This was considered an act of God: when he created a disease in a certain region, he mercifully added a curative in the same region (Boomgaard in prep.).

[47] Genetic engineering is also making its entrance into this field. At the moment researchers are attempting to develop a more effective myxoma virus. Such experiments are not without bio-risk, because the effects of a modified virus are unpredictable, and it has often proven impossible to control the spread of viruses (COGEM 2004).

[48] Vaccins may not work forever, since evolution always proceeds. An example of this is meningitis. Source: http://www.medscape.com/viewarticle/481484?src=mp. Researchers of the invasions of the microbe meningococcus W-135 associated with the *hajj*, listed in Box 20, note the following. After an epidemic of meningococcal disease by meningococcus of serogroup A in 1987 affecting 2000 pilgrims, Saudi authorities began requiring all pilgrims to be vaccinated against serogroups A and C before they could obtain a visa. The researchers presume that pilgrims can now be more easily infected by the W-135 strain (which actu-ally is a C strain in disguise). Although there is another vaccine that is also effective against W-135, it is feared that yet another meningococcus will appear, one that hides the C meningococcus or another dangerous serogroup under a misleading bacterial capsule (Hijmans 2003).

[49] In those areas where affordable antivirals were lacking (particularly in developing countries, where the majority of patients are found) this was partly compensated by hygiene measures like promotion and distribution of condoms as well as sterile hypodermic needles and syringes; disposable materials; and discouragement of breastfeeding in seropositive mothers. Other ways to prevent HIV infection include making surgical gloves widely available and testing donated blood for HIV infection. In recent years more affordable antiretroviral drugs have started to become available.

CHAPTER 5.
Impacts on nature, biodiversity and environmental quality

[50] Biodiversity can also be lost below species level, at the level of subspecies and ecotypes. Though sometimes quite significant, this effect will not be considered here.

[51] The vast majority of extinctions since AD 1500 have occurred on oceanic islands, but according to the IUCN, over the last 20 years, continental extinctions have become as common as island extinctions. However these figures apply to extinctions in general, not to extinctions due to bio-invasions. See: www.iucn.org/en/news/archive/2006/05/02_pr_red_list_en.htm

[52] On some small British islands, even the hedgehog caused damage after it had been introduced. Its diet includes seabird eggs (DEFRA 2003).

[53] *The Economist* and many liberal economists strongly believe in the benefits of *economic* competition. However, in *biological* competition following bio-invasions, the costs may far exceed the benefits, both to the economy and to biodiversity (at least for endemic species). That helps explain why countries like Australia, New Zealand and the US are on the one hand champions of free trade; on the other hand they have learned from past experience and now take great care to prevent bio-invasions. So the paradox is that economically, they are preaching the right of the strongest, while biologically, they champion the right of the weakest.

[54] In animal husbandry, animals kept to produce offspring for the export market are more and more kept under extremely stringent hygienic conditions. This

practice, though harmless to the environment, is questionable from the viewpoint of animal welfare. The animals no longer come into contact with the open air, do hardly build up natural resistance and generally lead extremely unnatural lives. Some pig breeders in the Netherlands even use caesarian section as a routine obstetric technique to produce "specific pathogen free" piglets for the export market.

55　The IPPC was established in 1951, partly in response to the invasion of the Colorado potato beetle in Europe. For treatment of wooden materials it allows a less harmful alternative to methyl bromide: heat treatment (Vermeulen & Kool 2006).

56　See http://www.winespectator.com/wine/daily/news/0,1145,2571,00.html.

CHAPTER 6.
Impacts on the economy and public health

57　Other sectors can also suffer damage from bio-globalisation. For instance, eco-tourism could suffer as invasive species move into more and more regions, whilst endemic species go extinct. Eco-tourists do not travel to Australia to see introduced rabbits, but to see endemic marsupials. The island of Guam has already lost much of its fauna attraction to eco-tourists. And if the dodo had not been driven to extinction, it could have been a gold mine for the tourist industry on Mauritius, comparable to the surviving Komodo dragon *Varanus komodoensis* on the Indonesian island of Komodo. Tourism is the world's largest economic sector, providing 10% of all jobs; eco-tourism has roughly a 20% share of that (Olden *et al.* 2005).

58　New Zealand faces a remarkable type of damage to agriculture. Since the mid-19th century the main source of nitrogen for dairy farms has been biological nitrogen fixation by white clover *Trifolium repens*. Following an invasion of the clover root weevil, this fixation has declined dramatically; dairy farmers are now forced to purchase much more artificial nitrogen fertilizer, at a cost of some $1.5 billion per year (Barlow & Goldson 2002).

59　MacIsaac *et al.* calculated the *potential* annual costs for damage and control of a *selection* of invasive exotic species in Canada (http://sgnis.org/publicat/macibail.htm). They arrived at:
- $5.3 – 14 billion for agriculture (by 6 exotic species)
- $7.7 – 20.1 billion for forestry (8 species)
- $299 – 776 million for aquatic systems (4 species).

That adds up to $13.3 - 24.9 billion annually. However, these costs are difficult to compare with actual costs calculated by Pimentel and by us.

60　Fleas (human, cat and dog fleas) are cosmopolitan commensals of humans and/or their pets. They are more attached to bodies than to houses and have probably been commensal for such a long time that it does no longer make sense in categorizing them as native or exotic. The same applies to lice, which have been commensals of Hominids for over half a million years. In America humans originally carried just one species of louse, *Pediculus humanus*. The Europeans brought with them two new subspecies and one new species, the head louse *P. h. capitis*, the human body louse *P. h. corporis* and the pubic louse *Phthirus pubis* (Wade 2004). Nowadays, all three of these (sub)species are cosmopolitan (Smart *et al.* 1948), but only the pubic louse can be said to be fully exotic in America.

61　The cholera epidemic of 1991 started in Peru, where the bacillus was either imported with ballast water or brought by El Niño. It cost Peru some $1 billion in loss of income from seafood exports and tourism (Bright 1998).

62　For the US and Brazil syphilis should perhaps not have been included because it may well be American in origin and may have occurred throughout the continent. But the cost of syphilis constitutes only a small fraction of the total – $18 million out of an annual total of $8.8 billion health costs for the two countries combined.

63　With such huge sums the question seems justified who is responsible for the costs. So far, the costs of flu are paid by the victims. But if we apply the Polluter Pays Principle, China would have to pay the huge bill (Van der Weijden & Van der Schans 2006).

64　BBC World 2005: *Gloomy estimate of bird flu costs*. See: http://news.bbc.co.uk/1/hi/world/asia-pacific/4414668.stm.

65　This category also includes some damage to agriculture, particularly damage inflicted by rats.

CHAPTER 7. The Netherlands and bio-globalisation

66　According to Rotterdam city ecologist Brekelmans, "Now and again a bird spider will enter with a shipload of fruit, and once a green iguana came in from South America. But these are not species that are going to settle here. The only species that I expect may spread outside (when winters are warm), particularly in the city where it is a degree or two warmer, is the flat back millipede *Oxidus gracilis*, orginally from East Asia. I happened across it in 2002 and 2003 at *Diergaarde Blijdorp* [Rotterdam zoo]. In Amsterdam it has also been found outside, in a compost heap at *Zorgvliet* cemetery. We can probably expect more species from the same group, as well as wood lice and centipedes." At least one exotic bird species has settled near Rotterdam harbour since 1994, presumably after hitchhiking on a ship: the house crow *Corvus splendens* from Southern Asia. It was found earlier around some seaports elsewhere and it is feared to become invasive.

[67] The coffee plant followed a most peculiar route. Initially, it was imported from Ethiopia to Arabia. The Dutch VOC brought it from Yemen to India, and from there (in 1690) to Indonesia. A cutting was sent from Indonesia to the Hortus (botanical garden) in Amsterdam. In 1715, the mayor of Amsterdam sent a cutting to the botanical gardens of Paris as a gift to King Louis XIV. From Paris, the plant travelled west to Martinique, where millions of coffee shrubs were soon growing (see also Box 48). By the year 1690, Paris counted 250 coffeehouses (hence the word "café"). And in 1734, J.S. Bach wrote his satirical *Kaffeekantate* on a dispute between a girl and her father, who thinks she drinks too much coffee.

[68] In a self-reinforcing chain of biological and societal factors, the Spanish initially attempted to use the Native Americans as slaves, for instance in Pizarro's Peru. For them, it was a logical step, since nearly half the South American subcontinent had been overrun by the Inca's and the majority of the population of the Inca Empire was already enslaved. The operation failed because pathogens introduced by the Europeans caused massive mortality among the Native Americans. This created a demand for African slaves, also logical insofar as that the slave trade was already common in Africa, mainly run by Arabic/Islamic traders. The Africans brought additional diseases to America, causing even more Native Americans to die and thereby creating even greater demand for African slaves. Previous to the 16th century, few European countries had engaged in any slave trade at all and the Spanish would never have started Transatlantic trading in slaves if sufficient numbers of healthy Native American labourers had been available to them.

[69] A recent arrival is the Atlantic croaker *Micropogonias undulatus*, a fish species originally found off the Atlantic coast of North America. Individuals have been found in the UK, in Belgium (1998) and in the Netherlands (2001). It was probably imported with ballast water (Buissink & Wolters undated).

[70] The list does not include plankton, parasites and epibiont species.

[71] According to Odé *et al.*, there were no invasions between AD 1500 and 1800. However, we do not consider this very credible.

[72] The terms *archaeophytes* and *neophytes* may give rise to confusion since they refer to the time of their settlement in the country, not to evolutionary age.

[73] Occasionally, an agricultural crop is introduced by accident. A possible example from the Netherlands is the cranberry *Vaccinium macrocarpon* from North America (habitually eaten by sailors to prevent scurvy). Legend has it that a barrel of cranberries was washed ashore on Terschelling Island in 1839 and been emptied in the dunes by a disappointed beachcomber. Be that as it may, the plant was identified in 1868 and settled on several parts of the island. This started commercial exploitation and cranberries are now also cultivated in some other regions in the Netherlands (Kalkman 2003). They are primarily used for making compote and wine.

[74] Denters examined herbs, not trees. Lever (2002) has compiled an inventory of exotic tree species in the South district of Amsterdam. The share of exotics in this group is even higher than one-third (see Box 74). But this high figure is even surpassed by the trees and shrubs of the Netherlands as a whole plus Flanders (Maes *et al.* 2006). Next to 114 indigenous species, they found over 3000 exotics! But not all of these species are fully naturalised.

[75] Remarkably, if we leave aside shore-weeds such as reed, the four swamping exotics - Japanese knotweed, giant knotweed, giant hogweed and Himalayan balsam - are among the tallest herbs in the Netherlands, the giant knotweed sometimes growing up to five meters. Giant hogweed also has very broad leaves. That parallels an Australian study, which found exotics to have on average relatively broad leaves, with invasive exotics having the broadest ones (Lake & Leishman 2004).

[76] The brown rat *Rattus norvegicus* displaced the black rat *R. rattus*, which however was also most likely an invasive species. On the British Isles, the black rat occurred from AD 100-400; it became extinct between 500-700, and then returned around 900. Today its distribution is limited to a few small islands (Williamson 1997). In the Netherlands, the black rat occurs on factory farms.

[77] The domestic cat is a borderline case. Its expansion is highly dependant on humans; there are only a handful of local autonomous populations. According to Nowak (1999) our domestic cat *Felix catus* has descended not from the native wild cat *F. silvestris silvestris* (though it is able to crossbreed with it) but from the subspecies *F. s. lybica*.

[78] According to Chapman (cited in Williamson 1997), damage done by Reeves' muntjac in the UK includes damage to vegetation by digging for subterranean

plant parts, as well as damage to forestry, orchards and gardening centres by feeding.

[79] Recently, even a marsupial was added to the fauna of Friesland, in the north of the Netherlands: the red-necked wallaby *Macropus rufogriseus*. This species originally escaped from a park and has managed to maintain its numbers in the past few years. An invasion seems unlikely.

[80] On the other hand, the (semi-exotic) domestic cat controls the rat at a local level.

[81] In 2003, 399,120 muskrats and 4847 coypus were caught (LCCM 2004).

[82] At most 1 pest among 9 invasive bird species – this neatly fits into Williamson's tens rule. But 3 to 4 pests (dependant on whether we include domestic cat or not) out of 12 mammals is significantly more than 10%, consistent with the higher figures mentioned by Jeschke & Strayer (2005).

[83] There may well be a relation between the high density of ticks in the *Waterleidingduinen* (Drinking Water Dunes) area near Amsterdam and the high density of fallow deer in that area.

[84] The term "passive distribution" to mean distribution through traffic and transport is somewhat ambiguous and misleading. The term is correct if seen from the perspective of the species, but ignores the active role that humans play in this kind of distribution.

[85] This inconsistency in the book could not be corrected due to a lack of sufficient data.

[86] An invasion of the Western corn rootworm could even affect a characteristic Dutch landscape. If the dairy farmer will have to alternate corn with grass, he may start growing corn on the land next to the stable, where he normally lets the cows graze. In that case he may decide to keep the cows in the stable all year round. That would accelerate an already ongoing trend that may lead to the gradual vanishing of the cow from the Dutch landscape.

[87] J. van Lenteren, cited in *De Volkskrant*, August 4th, 2004.

[88] In 1999, a natural enemy of the black cherry arrived from North America: the cherry fruit fly *Rhagoletis cingulata*. But this insect does not very much to control the black cherry and may rather damage cultivated cherry trees and related species.

[89] The calculated losses from animal diseases in the Netherlands are far below the range of 100-300% of prevention and control costs we have applied for agriculture and other industries. One reason for this is that - for fear of export bans - very contagious diseases (so-called list-A diseases) were not controlled by cheap methods such as vaccination, but by culling not only the few sick animals but also a large number of healthy animals. The significant damage involved is not included here in losses, but in control costs.

[90] For the sake of completeness we mention one exotic species which has had a beneficial effect on animal husbandry: the North American dump fly *Hydrotaea aenescens*, successfully introduced since 1950 to control house flies in stables.

[91] Among the proliferating geese there are also two native species that traditionally were winter visitors only, but recently have started summering and breeding in growing numbers: the barnacle goose *Branta leucopsis* and - to a lesser extent - white-fronted goose *Anser albifrons*. This is surprising from the viewpoint of climate. The (sub)tropical Egyptian goose now breeds next to two Arctic goose species!

[92] The caterpillars of the oak processionary moth *Thaumetopoea processionea* cause allergies, but strictly speaking this species is beyond the scope of this book because it was already present in the Netherlands in the 19th century and has now made a comeback, probably due to climate change.

[93] See the Global Invasive Species Program (GISP) by SCOPE, IUCN, UNEP and CABI: www.gisp.org

CHAPTER 8.
Lessons learned and actions needed

[94] A recent development in the Netherlands is the finding of the MRSA bacterium in pigs and pig farmers. Such resistant varieties can emerge locally or can be introduced. See: *Infectieziekten bulletin* 17: 272 (2006).

[95] For a nearly complete overview of pathways, see Appendix 5.

[96] In theory this should also apply to transportation from New Zealand and the temperate zones of Chili and Argentina to Europe. New Zealand does not have a very rich native species pool, but the much richer species pools of Chili and Argentina supposedly contain many potential pests for Europe.

[97] On 22 March 2006 a meaningful judgement was made by the Dutch *Raad van State* (Council of the State) on a case of three environmental organisations vs. the Ministry of Agriculture, Nature and Food Quality regarding permits issued for the introduction of mussel seed from the Irish and Celtic Sea into the Oosterschelde estuary. The Council decided there was a serious risk of introducing exotic species with unpredictable effects, which in view of the Precautionary Principle would not be acceptable in an area to be protected under the EU Habitats Directive. This risk was considered more important than the free trade of goods in the European market. The ministry had to nullify the permits. However, the European

Commission disagrees, judging that introduction of mussel seed is traditional practice. A ban would distort the common market. The Commission has requested the European Court to overrule the Dutch Council of State. So the conflict between free trade and 'ecological hygiene' continues.

[98] For an overview of the risks of deliberate introductions, see Van Dommelen 1996.

[99] We may be able to learn much from the fields of medical science, medical practice, public health organisations, phytopathology and veterinary science, in which diagnostic techniques have been developed to quickly detect key details. These techniques far surpass the diagnostic methods currently used in ecology.

[100] Regulation of ballast water transports can meet with opposition from the shipping sector. This is what happened in Canada, even after the introduction of the harmful oyster parasite MSX. See: http://www.ecologyaction.ca/news/103961478511996.html.

[101] The National Invasive Species Council (2001) lists four different ways in which the US has sometimes furthered bio-invasions in other countries: through development aid; military operations; food aid and international financing.

[102] It would be naive, however, to think that species introductions by bio-terrorists can be halted by conventional methods of detection, such as X-ray photography and sniffer dogs. Such introductions require only miniscule amounts of pathogens, which are virtually impossible to detect. Combating bio-terrorism requires different strategies and techniques.

[103] This is already being done by the Commission on Genetic Modification (COGEM) for genetically modified organisms and by the *College voor de Toelating van Bestrijdingsmiddelen* (CTB, Netherlands Board for the Authorisation of Pesticides) for biological pest control methods, but not yet for other species.

[104] On the US National Invasive Species Council, the Departments of the Interior, Agriculture, Commerce, Transportation and Defense, the State Department and the Department of the Treasury each have a seat, as does the US Environmental Protection Agency. The Council's primary task is to formulate an Invasive Species Management Plan. After 2001, president Bush brought all agencies monitoring the safety of in- and outgoing flows together under the new Department of Homeland Security. Belgium has a Biosecurity Council and New Zealand has a Biosecurity Law as well as a Minister for Biosecurity.

[105] Pigs have an even better sense of smell than dogs. In Southern Europe they are used to locate truffles that grow underground.

[106] This term is derived from the Hazard Analysis of Critical Control Points HACCP, which was developed to ensure the safety of the food chain – something that also frequently involves 'invasions' of microbes. The system may offer useful points of departure.

[107] An inspiring example of an innovation targeting bio-invasions is a technique developed by the Dutch company Greenship for decontaminating ballast water without chemicals. The technique uses filters followed by electrolysis. In addition, the company has developed a method that uses rapid DNA diagnostic techniques to determine whether purified water adheres to the future standards of the International Maritime Organization IMO (De Raat 2006).

Appendices

[108] Another factor may be that the volume of imported grain declined, since in the 1970s the EU became self-supporting in grain.

[109] At a 1990 symposium on "Reintroduction of species: rescue or face-lift for nature?" held in the Netherlands, a view taken on this issue was that it can be considered a good way of ecosystem management to (re)introduce key species for the functioning of the ecosystem (such as large herbivores). But without such an objective the information on habitat quality, as expressed in the species composition, is disturbed. In addition, endangered species can be reintroduced in special areas as an emergency measure. However, if the population starts declining again because mortality rates are not compensated for by birth rates and immigration, reintroduction may have to be repeated indefinitely (Verkaar & Van Wirdum 1991).

References

Anagnostakis, S.L. 1997: Chestnut and the introduction of chestnut blight. www.caes.state.ct.us/FactSheetFiles/PlantPathology/fspp008f.htm

Andersen, W.T. 1995: New role for CIA – Spying on biological invaders. http://news.ncmonline.com/news/view_article.html?article_id=c6225c32567474ccfba895e25fe3af43

Andeweg, R. & P. Florusse 2002: *Vreemde planten in Rotterdam*. Natuurmuseum Rotterdam, Rottterdam.

Aquasense 1998a: Ballast water. Overview of available data and estimation of possible risks. Ordered by the Ministry of Transport, Public Works and Water Management, North Sea Directorate. *Rapport* nr 98.1162. Amsterdam.

Aquasense 1998b: Risk assessment ballast water. Ecological and economic consequences, frequency analysis and measures. Ordered by the Dutch Ministry of Transport, Public Works and Water Management, North Sea Directorate. *Rapport* nr 98.1290. Amsterdam.

Bailey, L.H. 1928: The standard cyclopedia of horticulture. MacMillan & Co., London.

Bal, D. 2004: *Een overzicht van invasieve exoten in de soortgroepen Zoogdieren, Vogels, Reptielen, Amfibieën, Dagvlinders, Libellen en Mossen. Basisinformatie voor de Beleidsnota Invasieve Exoten. Expertisecentrum Ministerie van LNV*, Ede.

Ball, A.P. *et al.* 1979: Human botulism caused by Clostridium botulinum type E: the Birmingham outbreak. Quarterly Journal of Medicine 48: 473-91.

Barlow, N.D. & S.L. Goldson 2002: Alien invertebrates in New Zealand. In: Pimentel 2002.

Beijne Nierop, B.M. & T. Hakbijl 2002: *Ctenolepisma longicaudatum heeft ongemerkt bebouwd Nederland veroverd. Entomologische Berichten* 62: 34-42.

Benedictow, O.J. 2006: The Black Death 1346-1353. The complete history. The Boydell Press, Woodbridge.

Benedictus, R. 2004: *Het geld gaat naar IT, niet naar taxonomie*. Bionieuws 14 (20): 7.

Bergmans, W. & E. Blom (eds) 2001: Invasive plants and animals: is there a way out? Netherlands Committtee for IUCN, Amsterdam.

Bergmans, W. 2001: Concluding remarks and recommendations. In: Bergmans & Blom 2001.

Bibby, G. 1961: *Vierduizend jaar geleden. Het leven van 2000-1000 v. Chr.* Meulenhoff, Amsterdam.

Bieleman, J. 1992: *Geschiedenis van de landbouw in Nederland 1500-1950*. Boom, Meppel.

Bij de Vaate, A., K. Jazdzewski, H.A.M. Ketelaars, S. Gollasch & G. van der Velde 2002: Geographical patterns in range extension of Ponto-Caspian macroinvertebrate species in Europe. Can. J. Fish. Aquat. Sci. 59: 1159–74.

Bij de Vaate, B. 2004: *Invasieve aquatische exoten. Achtergrondnotitie van RIZA voor Beleidsnota Invasieve Exoten van LNV.*

Bilio, M. & U. Niermann 2004: Is the comb jelly really to blame for it all? *Mnemiopsis leidyi* and the ecological concerns about the Caspian Sea. Marine Ecology Progress Series 269: 173-83.

Biraben, Jean-Noel, 1980: An essay concerning mankind's evolution. Population, selected papers. Editions de l' I.N.E.D., Paris.

Blumenthal, D. 2005: Interrelated causes of plant invasion. Science 310: 243-44.

Boccaccio, G. (G.H. McWilliam, transl.) 1972: The Decameron. Penguin Books, London.

Bogaards, J.A., M.G.W. Dijkgraaf, J.M. Prins & F.de Wolf 2004: Changing direct costs of HIV treatment since the introduction of HAART in the Netherlands. *Jaarverslag 2003*. Stichting HIV Monitoring, Amsterdam.

Bokonon-Ganta, A.H. 2001: Introduction, economic importance and control of invasive pests in Africa; with special reference to the Mango Mealy Bug, Rastrococcus invadens Williams (Homoptera: Pseudococcidae). In: Bergmans & Blom (eds).

Bol, P. 1999a: *Gepokt. NRC-Handelsblad* May 29th.

Bol, P. 1999b: *Rhazes*. In: Kaandorp, C.J.E. & J.J.E. van Everdingen (eds) 1999: *Dokters van eeuwen. Twintig artsen uit twee millennia*. Belvédère, Overveen.

Bol, P. 2000: *Infecties met* Escherichia coli *type O157. Nederlands Tijdschrift voor Tandheelkunde* 107: 73-74.

Bol, P. 2001: *BSE en andere prionziekten. Nederlands Tijdschrift voor Tandheelkunde* 108: 72-73.

Bol, P. & A.E.M de Hollander 2001: *De decompartimentering van de wereld. Rapport voor minister J. Pronk van VROM.* RIVM, Bilthoven.

Bol, P., A.E.M. de Hollander & J. Melse 2001: *Onbeheersbare risico's van micro-organismen.* In:

Bouwstenen voor NMP4. Aanvulling op de Nationale Milieuverkenning 5. RIVM, Bilthoven.

Bol, P. 2002: *Importziekten 1 & 2. Nederlands Tijdschrift voor Tandheelkunde* 109: 321-22 & 366-67.

Bol, P. 2003a: *Influenza. Nederlands Tijdschrift voor Tandheelkunde* 110: 132-33.

Bol, P. 2003b: Severe Acute Respiratory Syndrome (SARS). *Nederlands Tijdschrift voor Tandheelkunde* 110: 335-37.

Bomford, M. & Q. Hart 2002: Non-indigenous vertebrates in Australia. In: Pimentel 2002.

Boomgaard, P. 2006: Syphilis, Gonorrhoea, Leprosy, and Yaws in the Indonesian Archipelago, 1500-1950. Paper.

Borio, G. 2006. The tobacco timeline. http://www.tobacco.org/History/Tobacco_History.html.

Bosman, W. 2003: *Het Rauwven, een "exotisch" ven in het beekdal van de Aa. RAVON* 15 (3): 33-36.

Bossenbroek, J.M., J. McNulty & R.P. Keller 2005: Can ecologists heat up the discussion on invasive species risk? Risk Analysis 25: 1595-97.

Bondt, N., C.J.A.M. de Bondt, G. Cotteleer, M. de Haan, H.H.W.J.M. Snegers & J.J. de Vlieger 2003: *Ontwikkelingen in de vleesindustrie tot 2007.* LEI, Den Haag.

Bradsher, K. & L.K. Altman 2004: A lethal virus stirs new fears of a pandemic. New York Times, October 16th.

Bright, C. 1998: Life out of bounds – Bioinvasion in a borderless world. W.W. Norton & Company, New York/London.

Brunet, R. 1901: *Maladies et insectes de la vigne. Librairie agricole de la maison rustique*, Paris.

Bryan, R.T. 1999: Alien species and emerging infectious diseases: past lessons and future implications. In: O.T. Sandlund, P.J. Schei & Å. Viken: Invasive species and biodiversity management. Kluwer Academic Publishers, Printed in the Netherlands.

Buissink, F. & M.H. Wolters undated: *Een vreemde eend in de bijt – Exoten in Nederland.* Niesje Wolters, Epe.

Caddy, J.F. & R.C. Griffits 1990: A perspective on recent fishery-related events in the Black Sea. General Fisheries Council for the Mediterranean Studies and Reviews 63: 43-71.

Canter Cremers - van der Does, E. 1965: *Het bad.* Van Dishoeck, Bussum.

Canyon, D., R. Speare, I. Naumann & K. Winkel 2002: Environmental and economic costs of invertebrate invasions in Australia. In: Pimentel 2002.

Cappuccino, N. & J.T. Arnason 2005: Novel chemistry of invasive plants. Biology Letters 2: 189-93.

Carlton, J.T. & J.B. Geller 1993: Ecological roulette: the global transport of nonindigenous marine organisms. Science 261: 78-82.

Chen, H. *et al.* 2005: Establishment of multiple sublineages of H5N1 influenza virus in Asia: Implications for pandemic control. Proc. Natl. Acad. Sci. 103: 2845-50.

Childs Kohn, G. 2001: Encyclopedia of Plague and Pestilence: From ancient times to the present. Checkmark Books, New York.

Chittenden, F.J. 1951: Dictionary of gardening. Clarendon Press, Oxford.

Clyti E., P. Couppie, C. Cazanave, F. Fouque, D. Sainte-Marie & R. Pradinaud 2003: *Traitement des myiases à Cochliomyia hominivorax par application locale d'ivermectine.* Use of ivermectin in Cochliomyia hominivorax's myiasis." Bull. Soc. Pathol. Exot. 96: 410-11.

COGEM 2004: *Ontwikkelingen in de biotechnologie - Achtergrondstudie bij de Trendanalyse Biotechnologie 2004.* CBD/CCMO/COGEM, Bilthoven.

Colautti, R. I. & H.J. MacIsaac 2004: A neutral terminology to define "invasive" species. Diversity and Distributions 10: 135-41.

Colautti, R. I., A. Ricciardi, I.A. Grogorovich & H.J. MacIsaac 2004: Is invasion success explained by the enemy release hypothesis? Ecology Letters 7: 721-33.

Convention on Biological Diversity 1992.

Cook, A., P. Weinstein & A. Woodward 2002: The impact of exotic insects in New Zealand. In: Pimentel 2002.

Crosby, A.W. 1972: The Columbian Exchange – Biological and cultural consequences of 1492. Greenwood, Westport, Conn.

Crosby, A.W. 1986 (reprint 2000): Ecological Imperialism - The biological expansion of Europe, 900-1900. Cambridge University Press, Cambridge.

Daehler, C.C. 2004: Review of Theodoropoulos 2003. Diversity and Distributions 10: 153-58.

Darwin, C. 1836 (reprint 1972): The voyage of the Beagle. Penguin, Harmondsworth.

Darwin, C. 1859 (reprint 1968): The origin of species by means of natural selection. Penguin, Harmondsworth.

Davis 2003: Biotic globalization: does competition from introduced species threaten biodiversity? BioScience 53: 481-89.

Davis, M.A., J.P. Grime & K. Thompson 2000: Fluctuating resources in plant communities: a general theory of invasibility. Journal of Ecology 88: 528-34.

Davison, N. 2005: The role of scientific discovery in the establishment of the first biological weapons programmes. University of Bradford, Bradford.

De Groote, H., O. Ajuonu, S. Attignon, R. Djoussou & P. Neuenschwander 2003: Economic impact of biological control of water hyacinth in Southern Benin. Ecological Economics 45: 105-17.

De Hollander, A.E.M. & P. Bol 2004: Friend and foe: health and the environment from an historical-epidemiological perspective. In: A.E.M. de Hollander. Assessing and evaluating the health impact of environmental exposures. RIVM, Bilthoven.

DEFRA 2003: Review of non-native species – Report of the Working Group. DEFRA, London.

De Nie, H. 1997: *Atlas van de Nederlandse Zoetwatervissen*. Media Publishing, Doetinchem: 1-151.

Den Hartog, C. & G. van der Velde 1987: Invasions by plants and animals into coastal, brackish and fresh waters of The Netherlands. Proc. Kon. Ned. Akad. Wet. C 90: 31-37.

Den Heijer, H. 1997: *Goud, ivoor en slaven. Scheepvaart en handel van de Tweede Westindische Compagnie, 1674-1740*. Walburg Pers, Zutphen.

Denters, T. 2004: *Stadsplanten – veldgids voor in de stad*. Fontaine Uitgevers, Abcoude.

De Raat, F. 2006: *Enge beestjes met het ballastwater weggooien, NRC Handelsblad* January 13th.

Der Spiegel 2004: *Anfällige Insulaner*. Nr 26: 131.

Davis, M.A. 2003: Biotic globalization: does competition from introduced species threaten biodiversity? Bioscience 53: 481-89.

Diamond, J. 1999: Guns, germs, and steel - The fate of human societies. W.W.Norton & Company, New York/London.

Dolin, R. 1998: Influenza. In: Fauci A.S., E. Braunwald, K. Isselbacher *et al.* (eds): Harrison's principles of internal medicine. McGraw-Hill, New York.

Drake, J.A., H.A. Mooney, F. Di Castri, R.H, Groves, F.J. Kruger, M. Rejmánek & M. Williamson (eds) 2001: Biological Invasions: A Global Perspective. SCOPE 37. John Wiley & Sons Ltd., Chichester.

Duyvendak, J.L. 1938: The true dates of the Chinese maritime expeditions in the early XV century. T'oung Pao 34: 341-412.

Duyvendak, J.L. 1949: China's Discovery of Africa. Probsthan, London.

EAAP 2003: After BSE – A future for the European livestock sector.

Elton, C.S. 1927: Animal ecology. Sidgwick and Jackson, London.

Elton, C.S. 1958 (reprint 2000): The ecology of invasions by animals and plants. The University of Chicago Press, Chicago.

Engelsman, M. Y. & O.L.M. Haenen 2003: *Ziekten en plagen van schelpdieren in Nederland*. Aquacultuur 2003: 22-25.

Invasive Species Pathways Team 2003: Final Report. USDA, Washington.

Fenner, F., D.A. Henderon, I. Arita, Z. Jezek, & I.D. Ladnyi 1988: Smallpox and its eradication. World Health Organization, Geneva.

Flinkenflögel, P.H. & P. Bol 1992: Rat-in-boots. On pesti-

lence, courage and impotence. In: J.J.E. van Everdingen (ed): The beast in man. Microbes and macrobes as intimate enemies. Part I. Belvédère, Overveen.

Gaastra, F. 2002: *De geschiedenis van de VOC*. Walburg Pers, Zutphen.

Geisler, W. 1994: AIDS: origin, spread and healing. Bipawo Verlag, Köln.

Genovesi, P. 2005: Eradications of invasive species in Europe: a review. Biological Invasions 7: 127-33.

Génsbøl, B. & J. Feilberg 2003: Plants and animals of Iceland. Edda Publishing Ltd., Reykjavik.

Goldschmidt, T. 1994: *Darwins hofvijver – een drama in het Victoriameer*. Prometheus, Amsterdam.

Gottlieb, M. & J. Weisman 1981: In: Morbidity and Mortality Weekly Report of the Centers of Disease Control, June 5.

Goudsmit, J. 2001: *Virus in water*. In: J.J.E. van Everdingen & J. Goudsmit (eds): *Gevaar van water, water in gevaar*. Belvédère/Medidact, Overveen/Alphen aan den Rijn.

Green, W. 2000: Biosecurity threats to indigenous biodiversity in New Zealand. www.pce.govt.nz/reports/all-reports/biosecurity_threats.pdf

Groeneveld, R. *Steriele mannen moeten reuzenpad bestrijden. BioNieuws* 30 June.

Gummer, H. 2006: Flightless ducks return home. World Birdwatch March: 13-15.

Gurevitch J. & D.K. Padilla 2004: Are invasive species a major cause of extinctions? Trends in Ecology and Evolution 19: 470-74.

Haas, G. 2001: *Entwicklung der Makro-invertebratengemeinschaft im hessischen Rhein- und Untermainabschnitt in den Jahren 1993 bis 1999*. Thesis J.W. Goethe-Universität, Frankfurt am Main.

Haas, G., M. Brunke & B. Streit 2003: Fast turnover in dominance of exotic species in the Rhine river determines biodiversity and ecosystem function: an affair between amphipods and mussels. In: Leppäkoski *et al.* 2003.

Haliday, T. 1978: Vanishing birds - Their natural history and conservation. Sidgwick & Jackson, London.

Harache, Y. 1992: Pacific salmon in Atlantic waters. ICES Marine Sciences Symposium 194: 31-55.

Henderson, L. 1995: Plant Invaders of Southern Africa. Plant Protection Research Institute, Agricultural Research Council, Pretoria.

Henderson, L. 2001: Alien weeds and invasive plants. A complete guide to declared weeds and invaders in South Africa. Agricultural Research Council, Pretoria.

Hengeveld, R. 2001: Invasion biology: from theory to practice. In: W. Bergmans & E. Blom 2001.

Heywood, V.H. 1989: Patterns, Extents, and Modes of Invasions by Terrestrial Plants. In: J. A. Drake *et al.* (eds) (2001).

Hijmans, A. 2003: AMC-Magazine, May.

HIV Monitoring Foundation 2006: Country Progress Report 2006 – Netherlands. Amsterdam.

Hofhuis, A., *et al.* 2006: *De ziekte van Lyme in Nederland tussen 1994 en 2005: Drievoudige toename van het aantal huisartsconsulten en verdubbeling van het aantal ziekenhuisopnames. Infectieziekten Bulletin* 17: 238-40.

Hofstede, R. & W.J. Wolff 2002: *De introductie van nieuwe soorten in de Nederlandse Noordzee. Een ecologische graadmeter voor klimaatverandering volgens de GONZ-systematiek.* Unpublished report RU Groningen, commissioned by the National Institute for Coast and Sea.

Huirne, R.B.M., M. Mourits, F. Tomassen, J.J. de Vlieger & T.A. Vogelzang 2002: *MKZ - verleden, heden en toekomst. Over de preventie en bestrijding van MKZ.* Landbouw-Economisch Instituut, Den Haag.

Hull, Th.G. 1955: Diseases transmitted from animals to man. CC Thomas, Springfield.

Huq, A., E.B. Small, P.A. West, M.I. Huq, R. Rahman & R.R. Colwell 1983. Ecological relationships between Vibrio cholerae and planktonic crustacean copepods. Applied Environmental Microbiology 45: 275-83.

Infectieziekten Bulletin 2003, nr 5: various articles on Lyme disease.

Insight Guide Zuid-Afrika 2002. Cambium, Zeewolde.

Invasive Species Pathways Team 2003: Final Report. USDA/APHIS. www.invasivespeciesinfo.gov/docs/toolkit/pathways.doc

ISSG undated: 100 of the world's worst invasive alien species. www.issg.org/database

Jeschke, J.M. & D.L. Strayer 2005: Invasion success of vertebrates in Europe and North America. Proc. Natl. Acad. Sci. U.S.A. Online 102: 7198-202.

Johnson, L.E. & D.K. Padilla 1996: Geographic spread of exotic species: ecological lessons and opportunities from the invasion of the zebra mussel Dreissena polymorpha. Biological Conservation 78: 23-33.

Kalkman, C. 2003: *Planten voor dagelijks gebruik – botanische achtergronden en toepassingen.* KNNV, Utrecht.

Karesh W.B., R.A. Cook, E.L. Bennett & J. Newcomb 2005: Wildlife trade and global disease emergence. Emerging Infectious Diseases. http://www.cdc.gov/ncidod/EID/vol11no07/05-0194.htm

Karesh, W.B. & R.A. Cook 2005: The human-animal link. Foreign Affairs, July/August.

Kearney, M.P. & B.S. Morton, 1970: The distribution of *Dreissena polymorpha* (Pallas) in Britain. Journal of Conchology 27: 97-100.

Kideys, A.E. 2002: Fall and rise of the Black Sea Ecosystem. Science 30: 1482-84.

Kideys, A.E., S. Ghasemi, D. Ghninejad, A. Roohi & S. Bagheri 2001. Strategy for combating Mnemiopsis in the Caspian waters of Iran. Report prepared for the Caspian Environment Programme. Baku, Azerbaijan.

Kilpatrick, A.M. *et al.* 2006: West Nile Virus epidemics in North America are driven by shifts in mosquito feeding behavior. PLoS Biology 4(4): e82.

Kolata, G. 2001: Flu. Pan Books, London.

Kouba, V. 2004: History of the screwworm (*Cochliomyia hominivorax*) eradication in the Eastern Hemisphere. Historia Medicinae Veterinariae 29: 43-53.

Kruger, A., A. Rech, X.Z. Su & E. Tannich 2001: Two cases of autochthonous Plasmodium falciparum malaria in Germany with evidence for local transmission by indigenous Anopheles plumbeus. Trop. Med. Int. Health. 2001: 983-85.

Lach, L., M.D. Picker, J.F. Colville, M.H. Allsopp & C.L. Griffiths 2002: Alien invertebrates in South Africa. In: Pimentel 2002.

Lake, J. & M. Leishman 2004: Invasion success of exotic plants in natural ecosystems: the role of disturbance, plant attributes and freedom form herbivores. Biological Conservation 117: 215-26.

Lange, R., P. Twisk, A. van Winden & A. van Diepenbeek 1994: *Zoogdieren van West-Europa.* KNNV-uitgeverij, Utrecht.

LCCM 2004: *Landelijk Jaarverslag 2003. Landelijke Coördinatiecommissie Muskusrattenbestrijding,* 's-Hertogenbosch.

LEI 2001: *Bestrijdingsscenario's voor Mond- en Klauwzeer.* Landbouw-Economisch Instituut, Den Haag.

LEI 2003a: *Ketenconsequenties van de uitbraak van vogelpest.* Landbouw-Economisch Instituut, Den Haag.

LEI 2003b: *Landbouw Economisch Bericht.* Landbouw-Economisch Instituut, Den Haag.

Leppäkoski, E., S. Gollasch & S. Olenin (eds) 2003: Invasive aquatic species of Europe. Distributions, Impacts and Management. Kluwer Academic Publishers, Dordrecht.

Levathes, L. 1994: When China ruled the seas. Oxford University Press, Oxford.

Lever, J. 2002: *Bomengids van Amsterdam-Zuid - Wandelend bomen leren kennen.* Ginkgo, Leiden.

Lieverse, R. & J.J.E. van Everdingen 2002: *Gekkekoeienziekte: de BSE voorbij?* Belvédère, Overveen.

Lobo Junior, M. 2002: Alien plant pathogens in Brazil. In: Pimentel 2002.

Longhurst, A. 1998: Ecological Geography of the Sea. Academic Press, San Diego.

Ludsin, S.A. & A.D. Wolfe 2001: Biological invasion theory: Darwin's contributions from The Origin of Species. Bioscience 51: 780-89.

Lyell, C. 1832: Principles of geology. Murray, London.

MacArthur, R.H. & E.O. Wilson 1967: The theory of island biogeography. Princeton University Press, Princeton.

MacIsaac, H.J., T.C. Robbins & M.A. Lewis 2002: Modelling ships' ballast water as invasion threats to the Great Lakes. Canadian Journal of Fisheries and Aquatic Science 59: 1245-56.

MacIsaac, H.J., J.V.M. Borbely, J.R. Muirhead & P.A. Graneiro 2004: Backcasting and forecasting biological invasions of inland lakes. Ecological Applications 14: 773-83.

Madl, P. & M. Yip 2003: Literature review on the aquarium strain of *Caulerpa taxifolia*. 31st BUFUS Newsletter, http://www.sbg.ac.at/ipk/avstudio/piero-fun.ct/ct-1

Maes, B. 2006: *Inheemse bomen en struiken in Nederland en Vlaanderen.* Uitgeverij Boom, Amsterdam.

Major, R.H. 1965: Classic descriptions of disease. Thomas, Springfield.

McEvedy, C. & R. Jones 1978: Atlas of World Population History. Facts on File, New York.

McKinney, M.L. & J.L. Lockwood 1999: Biotic homogenization: a few winners replacing many losers in the next mass extinction. TRENDS in Ecology and Evolution 14: 450-52.

McNeil, D.G. 2004: Wild birds, blamed for the spread of disease, often become its victims. The New York Times, 16 October.

McNeill, W.H. 1998: Plagues and Peoples. Anchor Books, New York.

McNeill, J.R. 1999: Ecology, epidemics, and empires: environmental exchange and the geopolitics of tropical America, 1600-1825. Environmental History 5: 175-84.

McNeill, J.R & W.H. McNeill 2003: The human web. W.W. Northon & Company, New York/London.

Menzies, G. 2003: 1421. The year China discovered America. William Morrow/HarperCollins, New York

Millett, J. 2006: Pacific Islands: a paradise lost? WorldBirdwatch, June: 21-24.

Monke, J. 2005: Agroterrorism: threats and preparedness. Congressional Research Service / The library of Congress. University Press the Pacific, Honolulu.

Muller, H.C.A. 1944. *Voorloopers en navolgers van Marco Polo.* Sijthof, Leiden.

Murray, C.J., A.D. Lopez, B. Chin, D. Feehan & K.H. Hill 2006: Estimation of potential global pandemic influenza mortality on the basis of vital registry data from the 1918–20 pandemic: a quantitative analysis. The Lancet 368: 2211-18.

National Invasive Species Council 2001: Management Plan: Meeting the Invasive Species Challenge. Washington.

Noordegraaf, L. & G. Valk. *De gave Gods - De pest in Holland vanaf de late Middeleeuwen.* Octave, Bergen.

Nowak, R.M. 1999: Walker's mammals of the world. The John Hopkins University Press, Baltimore/London.

NYFER 2002: *Door griep getroffen.* www.rivm.nl/vtvt/kompas/preventie/prevziekte/previnfectie/previnluenza/vaccgriep_effecten.htm

Odé, B., R. van der Meijden, C.L.G. Groen & W.L.M. Tamis 2003: *Invasieve neofyten in Nederland - een eerste verkenning van hun schadelijkheid. Rapport 2003.38.* Stichting FLORON, Leiden.

Okaichi T. 1989: Red tide problems in the Seto Inland Sea, Japan. In: T. Okaichi, D. M. Anderson, & T. Nemoto (eds): Red Tides - Biology, Environmental Science, and Toxicology. Amsterdam, Elsevier.

Office of the US Secretary of Defense 2001: Proliferation: threat and response. www.defenselink.mil.

Olden, J.D., M.E. Douglas & M.R. Douglas 2005: The human dimensions of biotic homogenization. Conservation Biology 19: 2036-38.

Pearson, D.E. & R.M. Callaway 2006: Biological control agents elevate hantavirus by subsidizing deer mouse populations. Ecology Letters 9: 443-50.

Pegtel, D. 2003: *Biogeografie en dispersie. De introductie van planten, dieren en micro-organismen. Collegedictaat RU Groningen.*

Perrings, C., K. Dehnen-Schmutz, J. Touza & M. Williamson. How to manage biological invasions under globalization. TRENDS in Ecology and Evolution 20: 212-15.

Petersen, K.S., K.L. Rasmussen & J. Heinemeier 1992: Clams before Columbus? Nature 359: 679.

Pimentel, D. (ed) 2002: Biological invasions - Economic and environmental costs of alien plant, animal, and microbe species. CRC Press, Boca Raton/London/New York, Wahington, D.C.

Pimentel, D. *et al.* 2002: Economic and environmental threats of alien plant, animal, and microbe invasions. In: Pimentel (ed) 2002.

Pimentel, D., L. Lach, R. Zuniga & D. Morrison 1999: Environmental and economic costs associated with non-indigenous species in the United States. Cornell University, Ithaca. http://www.news.cornell.edu/releases/Jan99/species_costs.html

Pimentel, D., R. Zuniga & D. Morrison 2005: Update on the environmental and economic costs associated with alien-invasive species in the United States. Ecological Economics 52: 273-78. *Note: we have used the corrected version kindly sent by Dr. Pimentel by e-mail on August 14th 2006.*

Pliny: *Naturalis historia.* Dutch translation 2004: *De wereld – Naturalis historia.* Athenaeum-Polak & Van Gennep, Amsterdam.

Pollitzer, R. 1959: Cholera. World Health Organization, Geneva.

Pratt H.D., S. Conant & J. Jeffrey 2005: The Hawaiian Honeycreepers: Drepanidinae (Bird Families of the World). Oxford University Press, Oxford.

Prenter, J. 2004: Roles of parasites in animal invasions. TRENDS in Ecology and Evolution 19: 385-90.

Prescott, W.H. 1901: History of the conquest of Mexico (1843) (J. Foster Kirk, ed). George Routledge and sons, London:

Prusiner S.B. 1996: Mad cows, cannibals, and prions. *Nederlandse Organisatie voor Wetenschappelijk Onderzoek*, The Hague.

PVE 2003: *Sectorinfo 2003. Productschappen voor Vee, Vlees en Eieren*, The Hague.

Quammen, D. 1996: The Song of the Dodo - Island biogeography in an age of extinctions. Simon & Schuster, New York.

Rajagopal, J., G. van der Velde, B.G.P. Paffen & A. bij de Vaate, 1998: Ecology and impact of the exotic amphipod, *Corophium curvispinum* Sars, 1895 (Crustacea: Amphipoda), in the river Rhine and Meuse. Reports of the project "Ecological Rehabilitation of Rivers Rhine and Meuse". No. 75-1998. RIZA, Lelystad.

Reemer, M. 2003: *Invasieve Arthropoda in Nederland: een eerste inventarisatie. Rapport in opdracht van de Plantenziektekundige Dienst. Stichting European Invertebrate Survey* – Nederland. Leiden.

Reumer, J. 2005: *De ontplofte aap. Opkomst en ondergang van de mens.* Contact, Amsterdam/Antwerpen.

Revenga, C. & Y. Kura 2003: Status and Trends of Biodiversity of Inland Water Ecosystems. Secretariat of the Convention of Biological Diversity, Montreal. Technical Series no. 11.

Ricciardi, A. & K.L. Rasmussen 1998: Predicting the identity and impact of future biological invaders. Canadian Journal of Fisheries and Aquatic Science: 1759-65.

RIVM 2002: *Volksgezondheid Toekomst Verkenning.* RIVM, Bilthoven.

RIVM 2003: *Nationaal Kompas Volksgezondheid.* www.rivm.nl/vtvt/kompas/preventie/prevziekte/ previnfectie/previnluenza/vaccgriep_effecten.htm

Roberts, Ch. & M. Cox 2003: Health & disease in Britain. From prehistory to the present day. Stroud, Sutton.

Ruddiman, W.F. 2003: The anthropogenic greenhouse era began thousands of years ago. Climate Change 61: 261-93.

Sala, O.E. *et al.* 2000: Global diversity scenarios for the year 2100. Science 287: 1170-74.

Sax, D.F. 2001 Latitudinal gradients and geographic ranges of exotic species: implications for biogeography. Journal of Biogeography 28: 129-50.

Sax, D.V. & S.D. Gaines 2003: Species diversity: from global decreases to local increases. TRENDS in Ecology and Evolution 18: 561-66.

Schilthuizen, M. 2006. *Tropenkolder in de ecologie.* Bionieuws 22 September.

Schwartz, M. 2004: Scientists question reports of massive ant supercolonies in California and Europe. http://www.eurekalert.org/pub_releases/2004-04/su-sqr040704.php

Shapiro, B. *et al.* 2004: Rise and fall of the Siberian steppe bison. Science 306: 1561-65.

Shaw, R.H. & L. A. Seiger 2002: Japanese Knotweed. In: R. Van Driesche *et al.* Biological Control of Invasive Plants in the Eastern United States. USDA Forest Service Publication FHTET-2002-04, Fort Collins, CO.

Shine, C., N. Williams & L. Gründling 2000: A guide to designing legal and institutional frameworks on alien invasive species. IUCN, Gland.

Simberloff, D. 2000: Foreword in: Elton (2000).

Smart J., K. Jordan & R.J. Whittick 1948: A handbook for the identification of insects of medical importance. British Museum, London.

Smit, H.S., A. Bij de Vaate, H.H. Reeders, E.H. van Nes & R. Noordhuis 1993: Colonisation, ecology and positive aspects of zebra mussels (Dreissena polymorpha) in the Netherlands. In: T.F. Nelapa & D.W Schloesser (eds): Zebra mussels: biology, impacts and control. Lewis Publishers, Boca Raton.

Snow J. 1854: The cholera near Golden Square, and at Deptford. Medical Times Gazette 9: 321-22.

SOVON Vogelonderzoek Nederland 2002: *Atlas van de Nederlandse broedvogels 1998-2000.* Nationaal Natuurhistorisch Museum Naturalis, KNNV Uitgeverij & European Invertebrate Survey-Nederland. Leiden.

Spielman, A. & M. D'Antonio 2001: Mosquito - A natural history of our persistent and most deadly foe. Faber and Faber, London.

Stirton, C.H. (ed) 1987: Plant invaders: beautiful but dangerous. Department of Nature and Environmental Conservation of the Cape Provincial Administration, Cape Town.

Takken, W., P.A. Kager & H.J. van der Kaay 1999: *Terugkeer van endemische malaria in Nederland uiterst onwaarschijnlijk. Nederlands Tijdschrift voor Geneeskunde* 143: 836-38.

Takken, W., R. Geene, W. Adam, T.H. Jetten & J.A. van der Velden 2002: Distribution and dynamics of larval populations of *Anopheles messeae* en *A. atroparvus* in the Delta of the Rivers Rhine and Meuse, The Netherlands. Ambio 32: 212-18.

Tamis, W., 2005: Changes in the flora of the Netherlands in the 20th century. Gorteria, suppl. 6: 1-233.

Taylor, L.H., S.M. Latham & M.E.J. Woolhouse 2001: Risk

factors for human diseases. Phil. Trans. Royal Soc. London 356: 983-89.

Theodoropoulos, D.I. 2003: Invasion biology: critique of a pseudoscience. Avvar Books, Blythe, California.

Thom, H.B. 1952: Journal of Jan van Riebeeck, Volume I, 1651-1655. A.A. Balkema, Cape Town/ Amsterdam 1952

Tien, N. & N. Dankers 2004: *Exoten in de Nederlandse kustwateren: Situatiebeschrijving en beheers- en onderzoeksvoorstellen. Achtergrondsrapport bij Beleidsnota Invasieve Exoten van LNV.* Alterra, Wageningen.

Tittizer, T, F. Schöll, M. Banning, A. Haybach & M. Schleuter 2000: *Aquatische Neozoen im Makrozoobenthos der Binnenwasserstrassen Deutschlands.* Lauterbornia 39: 1-72.

Tönjes, J. 2005: *Verdenking van bruinrot. Agrarisch Dagblad,* November 12th.

Torchin, M. E., K.D. Lafferty, A.P Dobson, V.J. Mckenzie & A.M. Kuris 2003: Introduced species and their missing parasites. Nature 421: 628-30.

Tsutsui, N.D. *et al.* 2000: Reduced genetic variation and the success of an invasive species. Proc. Natl. Acad. Sci. U.S.A. 97: 5948-53.

Tulp, A.S. 2006: *Mnemiopsis leidyi (Agassiz, 1865) (Ctenophora, Lobata) in de Waddenzee. Het Zeepaard* 66: 183-89.

Tye, A. 2005: Review of Theodoropoulos 2003. Biological Conservation 122: 505-07.

United Nations 1999: The World at Six Billions. http://www.un.org/esa/population/publications/sixbillion/sixbilpart1.pdf

Usher, M.B. 1988: Biological invasions of nature reserves: a search for generalizations. Biological Conservation 44: 119-35.

Van Dam, P.J.E.M. 2000: *De rol van de warande – geschiedenis van de inburgering van het konijn. Jaarboek voor ecologische geschiedenis. Vereniging voor Ecologische Geschiedenis. Academia Pers,* Gent.

Van der Heijden, C. 2001: *De aardappelmisoogsten (1845-1848). NRC-Handelsblad* September 26th.

Van der Meijden, R. 2005: *Heukels' Flora van Nederland.* Wolters-Noordhoff, Groningen/Houten.

Van der Putten, W. & F. Rienks 2004: *Amerikaanse vogelkers groeit ongeremd door bodemleven.* De Levende Natuur 105: 136-37.

Van der Velde, G., I. Nagelkerken, S. Rajagopal & A. Bij de Vaate 2003: Invasions by alien species in inland freshwater bodies in western Europe: the Rhine Delta. In: E. Leppäkoski *et al.* 2003.

Van der Weijden, W. 2001: *Landbouw kwetsbaar voor bioterreur. NRC Handelsblad* November 26th. English version: Agriculture vulnerable to terrorist attack. http://www.clm.nl/_engels/stichting_261101.html

Van der Weijden, W. 2003: *Rem nodig op biologische globalisering. NRC Handelsblad* September 9th. English version, somewhat more extensive: Biological globalisation needs to be checked. http://www.clm.nl/_engels/stichting_300903.html

Van der Weijden, W. & F. van der Schans 2006: *Griepgolven aanpakken bij bron in China. NRC Handelsblad* September 5th. English version: Time to fight flu at the source - in China. http://www.clm.nl./publicaties/data/fightfluchina.pdf

Van Dommelen, A. (ed) 1996: Coping with deliberate release – the limits of risk assessment. International Centre for Human and Public Affairs, Tilburg/Buenos Aires.

Van Hengstum, G. 2003: *Alleen genetische modificatie kan banaan redden. Bionieuws* 13 (2): 10.

Van Hoof, T.B., F.P.M. Bunnik, J.G.M. Waucomont, W.F. Kürschner & H. Visscher 2006: Forest-regrowth on medieval farmland after the Black Death pandemic – Implications for atmospheric CO_2 levels. Palaeogeography, Palaeoclimatology, Palaeoecology.

Van Kleef, H., G. van der Velde, R.S.E.W. Leuven & H. Esselink, in prep.: Ecological and economical consequences of Pumpkinseed sunfish (Lepomis gibbosus) invasions facilitated by nature management.

Van Lenteren, J.C., J. Woets, P. Grijpma, S.A. Ulenberg & O.P.J.M. Minkenberg 1987: Invasions of pest and beneficial insects in the Netherlands. In: W. Joenje, K. Bakker & L. Vlijm (eds.): The ecology of biological invasions. Proceedings of the Koninklijke Nederlandse Academie van Wetenschappen 90 (1): 51-58.

Van Riel, M.C., G. van der Velde, S. Rajagopal, S. Marguillier, F. Dehairs & A. Bij de Vaate 2006: Trophic relationships in the Rhine food web during invasion and after establishment of the Ponto-Caspian invader *Dikerogammarus villosus.* Hydrobiologia 565: 39-58.

Van Seventer, H.A. 1969: The disappearance of malaria in The Netherlands (thesis). Faddegon, Zwanenburg.

Van Strien, W. 2004: *Zijn de indringers nog te stoppen? Bionieuws* 14 (19): 5.

Van Wilgen, B.W., D.M. Richardson, D.C. le Maitre, C. Marais & D. Magdella 2002: The economic consequences of alien plant invasions: examples of impacts and approaches to sustainable management in South Africa. In: Pimentel 2002.

Veldhoen, L. 1997: *Mislukte rups uit Europa vreet Amerika kaal. De Standaard* 28 juli.

Verkaar, D. & G. van Wirdum 1991: *(Her-)introductie: redding of face-lift van de natuur? De Levende Natuur* 92: 196-200.

Vermeij, G.J. 1991: When biotas meet: understanding biotic interchange. Science 253: 1099-104.

Vermeij, G.J. 1996: An agenda for invasion biology.

Biological Conservation 78: 3-9.

Vermeulen, T., W. van der Weijden & A. Kool 2004: *Wereldwijde bloemenhandel is risico. Agrarisch Dagblad* July 5th.

Vermeulen, T. & A. Kool 2006: Phase out of methyl bromide as ISPM 15-treatment. CLM Research and Advice, Culemborg.

Vitousek, P.M., L.R. Walker, L.L. Loope, M. Rejmanek & R. Westbrooks 1977: Introduced species: a significant component of human-caused global change. Ecological Monographs 59: 247-65.

Volkskrant 2006a: *Amerikaanse ribkwallen vreten de Oostzee leeg.* December 2nd.

Volkskrant 2006b: *Hongerige roofkwal in Waddenzee.* December 16th.

Wade, N. 2004: The lessons of an ancient hominid and its lice. The New York Times, October 16th.

Wallace, A.R. 1876: The geographical distribution of animals. 2 vols. Macmillan and Co., London.

Wallace, A.R. 1880 (facsimile reprint 1998): Island Life. Prometheus Books, New York.

Wallerstein, I. 1976: The modern world-system, I. Capitalist agriculture and the origins of the European world-economy in the sixteenth century. Academic Press, New York.

Wallerstein, I. 1999: *Globalisering of transitie? Een lange-termijnvisie op de ontwikkelingsgang van het wereldsysteem.* Amsterdams Sociologisch Tijdschrift 26: 145-63.

Ward, N.B. 1842: On the growth of plants in closely glazed cases. J. Van Voorst, London.

Wassink, G. 2006:. *Nieuw geurtje helpt tegen mierenplaag.* Bionieuws 22 September.

Welcomme, R.L. 1992: A history of international introductions of inland aquatic species. ICES Marine Science Symposium 194: 3-14.

Werger, M.J.A. 1989: *Botanische tuinen. Een bijzondere wereld aan planten. Botanische tuinen Utrecht 1639-1889. Nu en in de toekomst.* Zomer en Keuning, Ede/Antwerpen.

Wetsteyn, L.P.M.J. & M. Vink 2001: Ballast water. An investigation into the presence of plankton organisms in the ballast water of ships arriving in Dutch ports, and the survival of these organisms in Dutch surface and port waters. On behalf of the North Sea Directorate. Report RIKZ/2001.026, Rijswijk.

White, P.C.L. & S. Harris 2002: Economic and environmental costs of alien vertebrate species in Britain. In: Pimentel 2002.

Wilcove, D.S., D. Rothstein, J. Dubow, A. Phillips & E. Losos 1998: Quantifying threats to imperiled species in the United States. BioScience 48: 607-15.

Williams, P.A. & S. Timmins 2002: Economic impacts of weeds in New Zealand. In: Pimentel 2002.

Williamson, M. 1997: Biological Invasions. Chapman & Hall, London.

Williamson, M. 2002: Alien plants in the British Isles. In: Pimentel 2002.

Wills, C. 1996: Plagues - Their origin, history and future. HarperCollins, London.

Wittenberg, R. (ed) 2005: An inventory of alien species and their threat to biodiversity and economy in Switzerland. The Swiss Agency for Environment, Forests and Landscape, Delémont.

Wolff, W.J. 1999: Exotic invaders of the meso-oligohaline zone of estuaries in The Netherlands: Why are there so many? *Helgolander Wissenschaftliche Meeresuntersuchungen* 52: 393-400.

Wolff, W.J. 2000: Causes of extirpations in the Wadden Sea, an estuarine area in the Netherlands. Conservation Biology 14: 876-85.

Wolff, W.J. 2005: Non-indigenous marine and estuarine species in the Netherlands. *Zoölogische Mededelingen* 79: 1-116.

World Birdwatch 2003: Campbell Island Teal can go home. Nr. 25 (3): 4.

World Birdwatch 2005: Mauritius highway plans shelved. Nr. 27 (4): 6.

Xie, Y., Z. Li, W.P. Gregg & D. Li 2001: Invasive species in China – An Overview. Biodiversity and Conservation 10: 1317-41.

Yanez, A. 1950: *Cronicas de la conquista. Ediciónes de la universidad nacional autonoma,* Mexico.

Zimmer, C. 2000: Parasite Rex – Inside the bizarre world of Nature's most dangerous creatures. The Free Press, New York.

Key web sites

Global Invasive Species Program: www.gisp.org
IUCN/ISSG: www.issg.org/database
Australia: www.marine.csiro.au/crimp/
UK: www.jncc.gov.uk/marine/dns/...
US: www.invasivespecies.org
US: www.invasive.org
US: www.cdc.gov./mmwr

Illustration Credits

Every effort has been made to trace and acknowledge all copyright holders. KNNV Publishing would like to apologize if any omissions have been made.

a=above, b=below, l=left, r=right

Photographers/illustrators
Scott Bauer, USDA Agricultural
  Research Service: page 97a.
Robert Beall, Flathead Valley Community
  College, Kalispell, Montana, US: 58l.
Gé van Beek: 29r, 147a.
Kees van Berkel/SAXIFRAGA: 125.
Stephan Biel, Robert Koch Institute, Berlin (courtesy
  H-W Ackermann, Dpt of Microbiology, Medical
  Faculty, Laval University, Quebec, Canada): 137.
Centre for Shellfish Research, Yerseke: 59r.
Stefan Claessens/KINA: 40.
Hans Dijkstra/Bureau voor Beeld: 86a.
Michael Dryden, Kansas State University, US: 153.
ENTOCARE CV: 61.
FloraHolland: 165.
J. Franssen/KINA: 53.
Yvonne Gooijer/CLM: 87 potato field, potatoes,
  maize field.
Francien de Groot/CLM: 8.
Cris Hake/ ANP: 37.
Wilmer Hendriks and Darryn Bruce: 75.
Martin Hill, Rhodes University, South Africa: 65.
T. Hill: 71r/a.
Luc Hoogenstein/SAXIFRAGA: 56.
Institute of Tropical Medicine,
  Antwerpen, Belgium: 14.
Chiel Jacobusse: 41, 77, 127, 146.
F. John: 141.
Martijn de Jonge/KINA: 149a.
L.F. Kamps: 149, small.
Thijs Keijzer: 102l.
Dick Klees/KINA: 18, 57, 152 r/a, r/b.
Lyubomir Klissurov: 60 l, r.
Nick Kontonicolas: 80.
Anton Kool/CLM: 87 wheat field, wheat.
Jan Kros/Rotterdam Zoo: 15.
Adrej Kunca, Narodne Lesnicke Centrum, Poland: 93
  small.

Rob Leewis: 52l, 54a, 58r, 59l, 167.
Rob Leewis/KINA: 100.
Willem Ligtvoet: 70.
Hanny van Megen, Wageningen University: 130.
R.Nolly/NHPA/Foto Natura: 20.
M. Otten: 144.
Re_Pop/KINA: 102r.
Reed Business Information: 93 large.
David Robinson. APHIS, NAS, US: 64.
Gordon Rodda, U.S. Geological Survey: 73 l, r.
Lies Rookhuizen, from: Pratt *et al.* 2005: 68.
Hans Schinkel: 32, 123.
Piet Spaans: 147b.
Alton N. Sparks, Jr., The University of Georgia: 97b.
Forest & Kim Starr (USGS): 72.
Edwin Sterckel: 101.
Stoelwinder/KINA: 29l, 149b.
USDA APHIS Archives
USDA Animal and Plant Health Inspection Service:
  54b.
USDA Forest Service - Northeastern Area Archives,
USDA Forest Service: 67.
U.S. Fish & Wildlife Service: 113.
U.S. Geological Survey Archives, U.S. Geological
  Survey: 63.
Luuk Wagter/KINA: 131.
Erik van Well, Francien de Groot and Erik de Bruin:
  22, 116.
Webbink/KINA: 149r.
Wouter van der Weijden: 86b, 87 maize, 122, 132
  large, 133 l, r/a, 158r.
West Coast Ballast Outreach Project: 168.
Fred van Wijk/KINA: 52.
Alex Wild, US: 42r.
Erik Wolbers: 31.

All graphics by Rob Leewis, unless stated otherwise.

Some photos were taken from the following
websites:
gtresearchnews.gatech.edu
www.apsnet.org
www.jenskleeman.de
www.nal.usda.gov
www.pixelfight.com.

Bio-invasions: definitions and their implications

Invasion biology is a young science. That is why we were constantly confronted with problems of definition and demarcation while writing this book. Often researchers use implicit concepts without realising what the implications are for their argument. Below we give some examples. The terminology will no doubt become more settled in coming years, but for now, we will give a tentative inventory of bottlenecks.

Time

One of the first questions rising is: where do we set the starting point for anthropogenic biological globalisation? Obviously, the earlier we set the starting point, the more species we have to regard as "exotic". In the end, every species on every location is exotic, unless it did evolve in that particular location from its predecessor.

The first option is the time when *Homo sapiens* started spreading from Africa, between 200,000 and 100.000 years ago ("Out of Africa 2", following "Out Africa 1" by earlier Hominids like *Homo ergaster* and *Homo erectus* between 2 and 1 million years ago). In that case we have to take into account the many cyclical invasions, extinctions, reintroductions etc. driven by the Ice Ages. The overall picture becomes extremely confusing, so this is not very useful as a cut-off point.

A more serious candidate is the *end of the last Ice Age* (9000 BC). The scene since that time has been dominated by invasions from the south as a result of global warming. Many species associated with human habitation were involved. Using this reference, most species in the northern temperate zone have to be considered exotic.

The *development of agriculture and towns* (between 8000 and 6000 BC) is another candidate. One complication here is that agriculture developed independently at different times in different parts of the world. Obviously introductions associated with agriculture and livestock husbandry play a key role. And aigain, if we use this reference period, countless species are to be considered exotics.

Another reasonable starting point is the period during which *cities developed into networks* and established transcontinental trading contacts. McNeill & McNeill (2003) call this "the human web". Here we also encounter the complication that no clear point in time can be identified. The networks of Chinese, Indian, Mesopotamian, Phoenician and Roman cities developed during a two-millennium period, from around 2000 BC until the 1st century AD (Bibby 1961). These represented the first steps

toward the globalisation of relations. This starting point is of particular value if the focus is on the spread of human infectious diseases (McNeill 1976, Bol & De Hollander 2001, De Hollander & Bol 2004).

A crucial phase begins in the early 15th century with the Chinese and Portuguese voyages of discovery (see Box 7). The Chinese soon quit but the Spanish join in. Following the "discovery" of America in AD 1492, a *global market* develops which begins to exert increasing influence over all the connected local economies (Wallerstein 1976, 1999). Transoceanic introductions of plants, animals and microbes, both intentional and unintentional, become increasingly important. The ultimate phase of biological globalisation starts in 1492, often simplified to "1500".

Starting with the *Industrial Revolution* (1780-1860), the number of movements of people and goods begins to increase rapidly, as do the distances covered. This accelerates the process of biological globalisation. Systematic colonialism, starting in the mid-19th century, accerelates the process even further.

A final, more pragmatic option is to choose a "simple" year - such as 1800, 1850, 1900 or 1950 - as a starting point; or a year in which systematic documentation started. For instance, Van der Meijden (2005) uses 1825 as the reference year for the indigenous flora of the Netherlands. In that year, the first more or less complete scientific Flora of the country was published.

Whether the dingo in Australia or the dog in Europe should be considered exotic species, entirely depends on the starting point chosen. The most appropriate starting point will depend on the objective of the inquiry, the questions raised and the data available. But obviously, the further back in time we choose our starting point, the more complications we have to face: ever more species about which we have ever less reliable data, and ever more fluctuations. In this book, we have employed several options concurrently; in some cases deliberately, in other cases because older data were absent.

Space

A bio-invasion is the settlement of a species in another area. In this book, we focus primarily on settlements across ancient distribution barriers. Barriers are present at every scale, but in invasion biology the barriers are geographical rather than ecological. Which types of barriers are relevant, then?

The first option is barriers between *continents*. Simple as it may be, this does not appear to have much relevance because not all continents are separated by distributional barriers. Europe and Asia actually constitute a single continent, where many species can move freely. Similarly, plenty of interchange is possible between North and South America and - to a lesser extent - between Africa and Eurasia.

A large distance in itself does not necessarily make an effective barrier. Wallace (1880) noted that there are more similarities between the fauna of the British Isles and Japan than between the fauna of Bali and Lombok - a distance of just 30 km! Conversely, distribution barriers can run right through continents – consider for example the Sahara, the Rocky Mountains, the Andes and the Pyrenees. And when a species from a region on a continent spreads to another region *within* that same continent – for instance the rabbit, which was introduced from the Iberian peninsula to other parts of Europe – would that not constitute a bio-invasion as well?

A better option are the borders between *biogeographical regions* such as the classical regions proposed by Sclater, which were adopted with slight modifications by Wallace (see Figure 2) and also used by Elton in 1958. Elton spoke of the "breakdown of the realms of Wallace" but that is problematic as well, since not all of Wallace's regions are separated by distribution barriers. In the Old World, interchange is possible between the Palaearctic and Oriental regions, as well as in the New World between the Nearctic and Neotropical regions. We can safely assume that most species able to survive on either side of such a biogeographical border have already crossed that border - though with little success.

Conversely, effective distribution barriers can exists *within* a region, such as continuous zones of forest, desert or tundra. For example: a number of species from East Asia have never reached Europe, and vice versa. This is not because of the great distance *per se*, but because of the (for those species) inhospitable climates and habitats in between. Wallace's regions are delimited partially by differences in climate, and partially by geographical barriers. The latter are most relevant to invasion biology because the field is primarily focused on invasions in which a species settles in an area that has a suitable climate and habitat, but was beyond its reach due to a physical geographical distribution barrier.

It therefore seems preferable to use the *old geographical barriers*: continuous areas that have had an unsuitable climate and/or habitat for an extended period of time. For example:

- For species inhabiting land, freshwater or brackish water: salt-water masses that did not become dry land during the Ice Ages. In addition to the huge

oceans, these barriers can also include ancient narrow sea passages, like the passage between Bali and Lombok.

- For species inhabiting salt water and brackish water: land masses such as continents. Again: the geographical mass does not have to be large as long as it is continuous, such as the Isthmus of Panama.
- *Within* land and water masses: continuous zones of climate and habitats unsuitable to the species in question (e.g. high mountain ranges, tropical seas or cold seas) or biomes (such as tropical rainforest, tundra, taiga and desert).

In other words, *bio-invasions usually do not involve the settlement of a species in a different climate zone but, rather, its settlement in a similar climate zone on the opposite side of a (ancient) distribution barrier.* Briefly: *settlement in a region with a similar climate but much different biota.* Such barriers partially coincide with the limits of classical biogeographical regions (see Figure 2) but also partially cut right through them.

The question remains what to do with bio-invasions across *relatively young barriers*, like the sea passages between continents and continental islands. Literally speaking, they are of course bio-invasions. However, the flora and fauna on opposite sides of such a sea passage are usually very similar, although the island will normally have fewer species. Therefore, such an introduction rarely has dramatic ecological consequences. For instance, the British Isles have a higher proportion of exotics than has Australia, but the ecological impact has been much less dramatic.

However, this is not a decisive argument against using the term "bio-invasion". Furthermore, there are notable exceptions, such as the introduction of a predator on a continental island where the predator had not yet arrived when the land bridge was flooded by the rising sea, or had arrived but later became extinct. In such cases introduction of the predator can cause massive damage to ground breeding birds, especially seabirds.

Another exception is continental islands that originated recently, such as the Wadden Islands in the Netherlands. An invasion of the fox on these islands would decimate large breeding colonies of gulls. In short, we cannot avoid calling such a phenomenon a "bio-invasion", but it will perhaps be subject of ecology rather than invasion biology.

Stage

In bio-invasions, we can differentiate between four stages of penetration:

1 Import of a species into a closed-off environment

(cage, aquarium, fenced-off enclosure etc.).

2 Introduction of a species into an open environment (by escape or release).

3 Settlement: the species reproduces and spreads spontaneously.

4 Development of a species into a pest.

At which stage do we speak of an invasion? If it is stage 1, then all exotic animals in zoos as well as all exotic species in horticultural gardens, private gardens and cities should be considered invasive. This is stretching the term "invasion" much too far. If it is stage 2, then every single escaped or released individual must be considered an invasion. This also makes little sense. It makes more sense to speak of an invasion from stage 3. That means we do *not* speak of an invasion only when the species has developed into a pest, as some authors do.

Independency

Should we consider it a bio-invasion when a species is introduced and manages to survive but does not reproduce and only maintains its numbers through repeated introduction? That question touches on the term "naturalisation": a species is not considered to be part of the indigenous flora or fauna until it demonstrates certain characteristics of reproduction under the local conditions, occurring continuously during at least P time units, at a minimum of Q locations. That is why lists of introduced species often include the *extent* to which the species has settled (e.g. "observed once"; "established locally"; "common"). For instance, in the Netherlands species are considered "naturalised" if they have been found at multiple sites and have also managed to spontaneously reproduce for at least three generations.

However, the fact that a species survives for decades is no proof of naturalisation. For example, more than 100 plant species in the Meuse valley were introduced with wool imports for a wool-washing factory in Belgium. Downstream in the Netherlands, many of these "wool adventives" settled, were continuously found for decades and thus seemed to be fully naturalised. However, after the factory had closed down, all but two of these species soon went extinct. Obviously, these populations had depended on regular influx of propagules. Likewise, many "naturalised" grain adventives near Rotterdam disappeared around 1970, possibly as a result of better purification of grain in the country of origin as well as in Europe (R. van der Meijden pers. comm.). [108]

If we cannot predict when the influx will stop, and if so, whether the species will go extinct, the pragmatic solution is to consider it a naturalised invader after some years.

Continuity

Some species that have disappeared from an area later spontaneously return or are deliberately reintroduced. Examples are the griffon vulture in the French Cevennes and the otter and the beaver in the Netherlands. [109] Can such a species be considered an exotic? If so, how much time will have to be passed between extinction and resettlement?

If the time span is short, the term "exotic" is obviously inappropriate. But the longer the time span, the more room there is for choice, debate and confusion. For instance, the rabbit was introduced in the Netherlands in the Middle Ages and has become fully naturalised. Whether we consider this time span long enough to call the species native, is a matter of choice.

A related question is: how to classify a species that had originally become extinct due to *natural* causes but was later reintroduced? Then we can consider it a new element, although the term "exotic" may be misplaced.

An almost hilarious example is the fallow deer in the Netherlands. It is fully naturalised in several parts of the country and is considered a native species by many. Others, however, argue that the species was introduced by the Romans and therefore must be called an exotic. Still others go even further back in time, arguing that the species already occurred in the region before the last Ice Age; so what the Romans in fact had done was reintroducing a native species!

The succession of extinction and return may even happen repeatedly and rapidly. This, for example, is the case for some human pathogens such as influenza, which may return within a decade's time in an only slightly modified strain. Whether we call such a species exotic or native is a matter of choice.

Taxonomy

Although a species can become extinct, it can return as a different taxonomic variety. This is what happens, for example, with influenza almost every year. In such cases it is preferable to speak of a new invasion.

The case is less clear-cut when a variant spreads, then crossbreeds with an indigenous variant. For instance, the domesticated cat was introduced by humans from the Middle East into Europe, where it crossbred with the indigenous wild cat. Consequently, the domestic cat is not an exotic species *pur sang* in Europe, though it is in America. Similarly, the dog is not a purely exotic species in America, since the Amerindians already had dogs before the Europeans introduced theirs. They can interbreed.

The economy of bio-invasions: dilemmas in calculating costs

When calculating the economic damage caused by bio-invasions, several methodological questions arise. A brief overview.

From which point in time do we start counting invasions?

In Appendix 1 we demonstrated that different starting points can be chosen for (anthropogenic) biological globalisation. That choice of course is of key importance for which species should be considered exotic and what the costs from invasive exotics are. If we are interested in economic cost, it is useless to choose a starting point in a period about which we have no economic data. Perhaps the oldest relevant starting point is around AD 1500, when the global market was created.

As we saw, there are other options:

- The mid-19th century, when all Western countries became involved in the Industrial Revolution and systematic colonization.
- An arbitrary year such as 1800, 1900 or 1950.

Obviously, the earlier we choose our starting point, the greater the accumulated costs from bio-invasions.

Which economic effects we should include in our calculations?

It is useful to differentiate between direct costs, indirect costs, direct damage and indirect damage. We can illustrate this with the 2001 foot-and-mouth crisis in the Netherlands (see Appendix 3):

- *Direct costs* were the costs of culling and destruction of animals and compensation for farmers.
- *Indirect costs* were the costs of cordoning-off livestock farms, policing transport bans etc., as well as the costs of prevention.
- *Direct economic damage* was caused by loss of turnover and export bans.
- *Indirect economic damage* - particularly to hotels, restaurants and cafés - was caused by the closing of nature areas, the cancellation of events and the ban on transport.

What about effects that were not quantified?

Following epidemics and natural disasters, no one is to provide a complete overview of all effects. There will always be some over- and underreporting, and the effects that are brought to light can not always be properly quantified. For example, there was not a single case of SARS in the Netherlands in 2003; nevertheless, there were indirect costs (public information campaigns and airport surveillance) as well as economic damage from decreased air traffic and tourism, among other things. Of course such indirect effects should be included as much as possible.

How to deal with effects that cannot be expressed in monetary terms?

Money in many cases is a rather poor way to express the broad palette of effects. Effects such as the disruption of societal traffic or collective shock and depression due to the culling and destruction of tens of thousands of animals or the removal of chickens from one's garden can hardly be translated into a monetary value. The same goes for disease and mortality.

The only effects that can be expressed in terms of money are the cost of treatment for disease and the loss of working days for sick leave. In the latter case, the illness of a child or pensioner has less value than the infirmity of an employee – something which is defensible from an economic point of view, but hardly from a humane perspective.

In addition, damage to the environment and nature can be caused not only by the invasive species, but also by efforts to combat it. This also is difficult to express as money, but must not be ignored.

Economists have developed several methods to express multiple values and effects in monetary terms, including contingency valuation and hedonic pricing. Such methods attempt to measure, for instance, how much money individuals are willing to pay in order to retain certain values. In the US in particular, such methods are often used in legal cases.

An example here is biodiversity: Pimentel *et al.* (1999) calculated the monetary value of the birds killed by the 63 million domestic cats and 30 million feral cats in the US. Based on, among other things, the amount that birdwatchers pay per bird and the amount that hunters pay to hunt, they arrived at a figure of $30 per bird. Adding to that the number of fish killed by cats, the total estimated amount was $14 billion in damage per year. In an update Pimentel *et al.* (2005) go even further, estimating damage worth $17 billion just for birds killed by feral cats.

Such calculations will not be readily accepted by Europeans, even disregarding the fact that domestic cats are only semi-exotic in Europe. It seems preferable to supplement calculations of financial values with information on the loss of other values, such as human life, well-being, biodiversity, expressed in appropriate terms.

Would there be other costs if the species had not settled?

After determining the damage caused by an invasive species, we must pose the question: what damage would have been done had the species *not* settled? In some instances, such a 'what if?' question is not very difficult to answer, for example in the case of the muskrat in the Netherlands and the zebra mussel in the Great Lakes of North America. They occupy a vacant niche and are also recent settlers, so we do know the baseline situation.

In most other instances, the question is much more difficult to answer. The brown rat, for example, has caused a lot of damage in Europe. But had it not been there, its predecessor, the black rat, would certainly also have caused a considerable amount of damage. To give another example: in Portugal, California and elsewhere, frequent fires have occurred in plantations of introduced trees like Eucalyptus. If instead indigenous tree species had been used, how many fires would have occurred, and how much damage would have been involved? Quantifying such damage in some cases cannot be more than an educated guess.

What is the effect of predation on the population size of its prey?

If a predator consumes 5% of the available prey animals, then it is often simply assumed that the predator thereby reduces the population by 5%. But this will rarely be the case. Cats, for instance, feed primarily on young birds, many of which will not live beyond their first year in any case. In addition, sick animals are the primary targets of many predators; the result is positive selection. Furthermore, predation can limit the spread of an infectious disease to other individuals. For instance, foxes often take rabbits affected by myxomatosis. It is not easy to integrate such considerations into calculations of the economic costs from invasions.

How does one value the hunt and consumption of an invasive species?

In some cases the problem of a species invasion can be turned into an opportunity by "harvesting" the species. Examples are the consumption of the muskrat in Belgium (where it is commonly known as *waterkonijn*, or water rabbit) and the Chinese mitten crab in the Netherlands. The benefits can then be partially expressed in monetary terms (e.g. hunting rights) and must be subtracted from the calculated amount of damage caused by the species in question.

Should we include the costs of prevention in our calculations?

Bio-invasions can generate costs even if they do *not* occur: prevention costs. For instance:

- Measures at border checkpoints, including quarantine, to prevent pathogens and other harmful invaders from entering a country.
- Hygiene measures in hospitals against the entry of exotic bacteria. For instance, Dutch hospitals apply very stringent hygiene to prevent import of the MRSA bacterium.
- Vaccination of the population and livestock against pathogens that have disappeared from a country but may return.
- Production and storage of strategic stocks of antibiotics, antiviral drugs and vaccines in case border checks fail. For instance, the Netherlands has a national stockpile of cowpox vaccine.

Prevention costs can be considerable but as the old adage goes, "prevention is better than cure." To give one example: BSE. Partly because the Netherlands did not exercise enough care when importing livestock and feedstuff from the UK, every slaughtered cow is still being tested for BSE, at considerable costs for the farmers.

One advantage of border controls and hygiene is that they offer protection against a broad spectrum of invasive pathogens.

Obviously, the costs of preventive measures should be included in calculations of costs from bio-invasions. But that has its limits. It is no longer reasonable to attribute (part of) the costs of water and sewage costs to the cholera bacterium. Such infrastructure has become an integral part of civilised life, at least in the affluent countries.

How to deal with opportunity costs?

All labour and costs incurred in prevention and control of bio-invasions could also have been used for other activities that generate money or other benefits. Economists refer to this as "opportunity costs." For example: in the Netherlands, police officers were deployed to enforce transport bans during the foot-and-mouth disease crisis; these policemen would normally have been deployed for other purposes during that period, such as crime prevention. So it is not an odd assumption to say that an outbreak of a highly infectious animal disease may increase the level of crime for the time being. Economists have developed methods to estimate such costs.

Economic effects of invasive animal diseases in the Netherlands

Carin Rougoor PhD
CLM Research and Advice

This appendix provides further insight into the calculations of the costs from invasive animal diseases listed in Table 10. It can be seen as an exercise in calculating the incredibly complex economic impact of a bio-invasion.
The appendix includes:
1 The basis for our calculations
2 The costs from BSE
3 The costs from foot-and-mouth disease
4 The costs from classical swine fever
5 The costs from bird flu
6 Summary.

The basis for our calculations

Estimates of the economic damage caused by animal diseases are based on the following assumptions:
1 An outbreak of a highly contagious disease such as foot-and-mouth disease causes a large amount of damage in the year of the outbreak; in other years there is no damage because there is no outbreak. To estimate the average costs over multiple years we have adopted the following approach: the total costs from large-scale outbreaks like foot-and-mouth disease, swine fever, bird flu and BSE in the rather exceptional period 1995 - 2004 will be considered the *upper limit* of the calculated costs during a period of 10 years (i.e., 10% of that as the upper limit of the annual costs). So the assumption is that the number of outbreaks of contagious animal diseases and their effects was above average in the 10-year period. We will take a *quarter* of these costs as the *lower limit* for our calculations.
2 Indirect (and external) *benefits* will be left out of consideration. With this we mean the benefits to other sectors (e.g. the benefit to animal culling and destruction companies when animals are destroyed to eradicate a disease, or the benefit to other livestock sectors and the meat-alternatives producing industry. Such "one man's meat is another man's poison" effects (see Box 81) will be left out of consideration because they obscure the real damage of outbreaks.
3 We will limit ourselves to the costs to the *Dutch* economy; damage in other countries will be left out of consideration.
4 *Attributing* certain costs to a disease is dependent on the selected baseline situation. For example, the costs from the ban on meat and bone meal were attributed to BSE because this situation had arisen as a result of a BSE outbreak in the Netherlands. These costs make up a significant proportion of the total costs from BSE. This choice is rather arbitrary, because refraining from feeding ruminants with meat and bone meal can also be seen as normal practice, since ruminants are not carnivores.
5 A variety of hygienic measures are implemented on farms, in part to prevent disease from entering or leaving. This however is not the only goal; working hygienically is also necessary to ensure milk quality and animal welfare, for example. In addition, the costs cannot be attributed to a single infectious disease. We therefore do not include these costs in our overview.
These figures are based, as far as possible, on research by others. Where this was not available, we have made our own calculations.

The costs from BSE

The costs from BSE in the EU

The EAAP (2003) has calculated the costs from BSE on a European scale:
- A loss in value of €55 per animal (because offal cannot be used any more).
- A cost of €21 per animal for removal of meat and bone meal.
- A cost of €25-50 for BSE testing.

The total costs thereby add up to over €100 per animal for tested animals and €75 per animal for untested animals. This amounts to €2.8 billion per year for all 15 EU countries taken together. All estimated future costs included, the total costs come to €92 billion (this is the net present value [NPV]. Future costs will be automatically calculated into the total).

The costs from BSE in the Netherlands

Prevention

To prevent BSE, the following measures have been implemented in the Netherlands:

1 Since 1994, meat and bone meal may no longer be used to feed cows, sheep and goats, and the requirements for the production of meat and bone meal have become more stringent (€20 minutes at 133°C).
2 Brains, marrow, eyes, tonsils and parts of the intestines of cows, sheep and goats may no longer be consumed (effective from October 2000).
3 The BSE-monitoring programme (since January 2001) provides for the testing of slaughter material in slaughter-houses. A *post-mortem* test must be performed on:
   - Animals for emergency slaughter or animals that show signs of any disease and are over 24 months old.
   - All cows over 30 months in age that have been slaughtered in the normal fashion for human consumption (note: the age limit may be raised to 42 months; see *Agrarisch Dagblad*, December 31st, 2004).
   - Cows that have died on farms and are over 24 months old.

In practice, this means most animals used in dairy farming must be tested for BSE, while many animals bred for the meat industry do not need to be tested.

Damage and losses

When a case of BSE has been found, we can differentiate between direct and indirect costs:

Direct costs:
4 Culling and destruction of diseased animals.
5 Culling and destruction of other animals on the same farm as well as family members of the diseased animals.
6 Loss in income by farms struck by BSE, from a lack of animals following livestock culling and from acquiring new livestock.

Indirect costs
7 Effects on export: the export of beef dropped during the BSE crisis. Major beef-importing countries like Egypt and Iran banned beef from EU countries in 2001. In addition, the export of beef from the Netherlands to other EU countries came under pressure, because of the decrease in consumer demand (due in part to BSE) as well as a greater preference for domestically produced meat. The export of veal to Germany also came under pressure (dropping 27% from 2000 to 2001). It has been estimated that during 1999 and 2000, as a consequence of BSE:
   - The export of livestock fell by 2%.
   - The export of beef fell by 10%.
   - The export of veal fell by 6%.
   - The export prices of livestock, beef and veal fell by 3% (Huirne *et al.* 2002).
8 Domestically, the sector suffered damage to its reputation, which may have led to a decline in turnover. Huirne *et al.* (2002) have estimated that beef consumption fell by 1% and veal consumption by 4% due to BSE. Besides, the reputation damage may result in a weaker social status of the sector. However, the economic effects of this are hard to estimate.
9 Possible effects for other sectors: the poultry industry initially profited from the problems in the livestock industry.

However, as previously stated such effects will be left out of consideration here.
10 Follow-up damage due to humans contracting Creutzfeldt-Jakob disease. This has so far happened only twice in
the Netherlands.

<u>Further elaboration</u>

Re 1, 2, and 3: meat and bone meal and other offal, and testing

Every year some 656,000 cows are slaughtered in the Netherlands (LEI 2003b). Since January 1st, 2001, some of these are tested for BSE, and others are not. According to the Ministry of Agriculture, Nature & Food Quality, over half a million tests are performed every year. Previous to 2001, only cows showing symptoms of BSE were tested; in addition, a random sample of dead cows was also tested. According to PVE (2003) the costs of testing for BSE and removing high-risk material (since October 2002) in the Netherlands are between €35 and €160, depending on how efficiently the cows are slaughtered. Previous to 2001, only bone meal needed to be removed. The costs are estimated at €21 per animal (EAAP 2003). This means that the costs for the period 1995 - 2004 add up to:

Loss of value of the animals: 656,000 animals x €55 x 4.25 years = €360 million
BSE testing and removal of high-risk material: 500,000 animals x (€35-160) x 4 years = €70-320 million
Removal of bone meal: 656,000 animals x €21 x 6 years = €83 million

Total costs in the period 1995 – 2004: €513–763 million

Note: the ban on the use of meat and bone meal is under discussion. When this ban will be lifted or relaxed, the costs will drop.

Re 4. Culling and destroying diseased animals, and:
Re 5. Culling and destroying other animals on the affected farm and family members of the diseased animals.

In the period 1997 - 2004, there were 77 reported cases of BSE in the Netherlands. That is an average of 10 cases per year. Most cases (24) were in 2002; in 2004 the number had dropped to only 6 reported cases in that year.
Up until March 2003, all animals on a farm where a BSE-infected cow was found were destroyed; each case of BSE caused an additional 107 other animals to be destroyed, on average. From April 2003, only "relatives" of the BSE-infected animal are destroyed and the average had dropped to just 38 other animals destroyed per BSE case.
We can calculate the damage per animal based on the cost of replacing dairy cows and calves. An adult cow has a replacement value of about €1000. A calf has a replacement value of between €230-800, depending on its age (*KWIN-Veehouderij*). Here we will assume an average value of €750.
In the period 1997–2004, a total of 6851 animals were culled and destroyed in the Netherlands. At a value of €750 per animal, that amounts to €5 million.

Re 6. Loss of income by BSE-affected farms following animal culling and destruction, due to loss of livestock and the cost of acquiring new livestock

We assume that it takes a farm three months to replenish its livestock. Huirne *et al.* (2002) assumed €2.25 per dairy cow per day (including calves) for the costs from missing livestock.
Assuming that dairy cows made up around 4000 of the 6851 destroyed animals, the total costs from missing livestock amount to:

4000 animals x 90 days x €2.25 = €0.81 million

Re 7. Decreased export of meat and livestock and drop in prices

The Netherlands exports some 169,000 tons of beef annually (Bondt *et al.* 2003). In 2000, there was a drop of 6% in the export of veal and a drop of 10% in the export of beef. It is not clear whether this effect was temporary or permanent. The lower limit for our calculations will be that the slump in the export lasted just one year; our upper limit will be that the 10% decline in exports was permanent. The price of beef from slaughtered dairy cows is about €1.50 per kilo (KWIN-Veehouderij). Cows and calves raised for meat production fetch a higher price. For our calculations we will assume an average meat price of €1.75 per kilogram.

Lower limit: 16,900 tons of meat x €1.75 loss in turnover = €29.6 million loss in turnover
Upper limit: 16,900 tons of meat x €1.75 loss in turnover x 5 years = €147.9 million

The export of livestock fell by 2% while the price of livestock dropped by 3% (Huirne *et al.* 2002). The total export is some 40,000 animals per year. A slaughtered dairy cow fetches approximately €465; livestock raised for meat production fetches €500 (*KWIN-Veehouderij*). Based on the same assumptions as we chose for meat export, the loss in turnover is:

Lower limit: 40,000 animals x (3% of €500) + 800 animals x €500 =   **€1 million**
Upper limit: 5 times this loss in turnover =   **€5 million**

Re 8. Domestic drop in beef consumption

The consumption of beef in the Netherlands amounts to 287,000 tons. At least 80,000 tons of this is imported (Bondt *et al.* 2003). The costs of a 1% decrease in consumption (in a minimum of 1 year, up to a maximum of 5 years) are:

Lower limit: 2000 tons x €1.75   **€3.5 million**
Upper limit: 5 times this loss in turnover =   **€17.5 million**

Summary of the costs from BSE in the period 1995 – 2004 (in millions of euros)

Costs of prevention and control:	
Removal of offal, removal of meat & bone meal and testing	513-763
Destroyed animals	5
Loss of income by BSE-affected farms	1
Loss in turnover:	
Drop in meat exports	30-148
Drop in livestock exports, at lower prices	1-5
Drop in beef consumption in the Netherlands	4-18
Total	**€554–940 million**

The costs from foot-and-mouth disease

In 2001 there was a major outbreak of foot-and-mouth disease in the Netherlands. Huirne *et al.* (2002) analysed the economic effects of this outbreak:

Costs to the agricultural sector and suppliers (in millions of euros):

Loss of farm income (due to price changes and missing livestock; the costs of culling and destroying animals are not included because they were reimbursed by the *Diergezondheidsfonds* [DGF, Animal Health Fund])	230
Contribution of the agricultural sector to the DGF	120
Extra costs for animal feed suppliers (decontamination of vehicles, changed routes etc.)	15

Costs to the government (in millions of euros):

Costs to the Ministry of Agriculture, Nature & Food Quality (including the General Inspection Service AID) and lower-level authorities	67

Damage to the agricultural sector (in millions of euros):

Loss of income livestock transportation, trade, markets and slaughter	183
Loss of income and costs dairy processing and livestock improvement	40
Loss of income agricultural services (contract work, advisory service)	10

Damage to other sectors (in millions of euros):

Tourism, recreation, retail, construction and cargo transport (loss of income)	208
Costs to nature and landscape conservationists	1

Total (in millions of euros)	**874**

The costs from classical swine fever

A major outbreak of classical swine fever struck the Netherlands in 1997. According to Wageningen University and Research Centre (WUR), this outbreak cost over **3 billion guilders (€1.4 billion)**. The specific costs were (in millions of euros):

Animals culled on 362 contaminated farms and preventively culled on an additional 881 farms	226
4.4 Million animals that were bought up for reasons of welfare and 1.4 million piglets which received a lethal injection	454
Transport restrictions, missing livestock following culling, feed industry, artificial insemination, trade, transport and slaughter	681

These are the costs of eradicating the disease. The EU paid for 35% of these costs, the Dutch government paid for 20%, pig farmers paid 25% and other parties in the sector (slaughter houses, trade, transport, animal feed industry, artificial insemination stations) paid the remaining 20%.

No exact figures are available for the loss in turnover or profits due to swine fever. Farms not directly affected had a financial advantage: after swine fever broke out, prices rapidly jumped over 20% (*Centraal Bureau voor de Statistiek*, CBS, National Statistics Agency). On the other hand, consumers may have started buying less pork meat. There was also a negative effect on the export of pork: the US, for example, closed its borders to Dutch pork between 1997 and 2003. According to the *Productschap Vee en Vlees* (PVE, Product Board for Livestock and Meat), the loss in income by pig farmers was 200 million guilders (€90 million) just for the period up to September 2001. For the period as a whole (1997 - 2003) the loss of income was an estimated €135 million.

In 1997, there were a total of 15.2 million pigs on Dutch farms. As a result of swine fever, this number dropped to 13.4 million in 1998. Though recovering slightly in 1999, when pigs totalled 13.6 million, the number subsequently dropped back down to 11.3 million in 2004. As the lower limit for our calculations we will assume that the decline in pigs would have occurred even if the swine fever outbreak had not happened. Our upper limit will be based on the assumption that the drop of 1.5 million in the total number of animals (hogs, piglets and sows combined) is a permanent effect of the swine fever outbreak, with the result that 3 million fewer pigs are produced per year. The surplus for each pig is about €22. On an annual basis this represents a loss of €66 million. For the total period 1998 - 2004 this amounts to **€462 million**.

This figure is the maximum possible amount of damage; the closing of US borders to Dutch pork meat can, after all, be offset by the reduction in the number of pigs. The minimum possible amount of damage consists only of the damage caused by the closing of US borders: **€135 million**.

The total costs thus amount to between **€1.5 and €1.9 billion**, equivalent to **0.5% of the 1997 GDP**.

The costs from bird flu

According to the *Landbouw Economisch Instituut* ((LEI, Agricultural Economics Research Institute), the costs from the bird flu outbreak in 2003 were at least €750 million:

Cost of culling and destroying animals	**€270 million**
Loss of income from eggs and poultry meat (due to Lack of poultry animals after culling etc.)	**€500 million**

These costs were strictly for the purpose of eradicating bird flu. Not included are the costs from diseases among 89 humans, one of which died.

Half of the costs of animal culling and destruction were reimbursed by the EU. Half of the pre-crisis export level of eggs (over 7 billion eggs per year) was expected to be permanently lost (*De Volkskrant*, July 4th, 2003). According to the Product Board for Livestock and Meat (PVE), the production of eggs was about 9.5 billion units in 2004. This is an increase of 32% from the previous year, but a drop of 4% from 2002 (*Sectorinfo Pluimvee en Eieren* 10, October 26th, 2004). As our upper limit we will assume that this decrease in production is entirely attributable to the outbreak of bird flu. Over a span of two years, this represents a drop in total egg production of some 3 billion eggs. The producers' price of cage eggs is about 4.5 eurocent *(KWIN-Veehouderij)*. Using these figures, the costs amount to **€135 million**.

The PVE has estimated that Dutch production of poultry meat amounted to 596 million kilo of slaughtered animals in 2004. That is a drop of 15% from 2002, but a jump of 11% from 2003. The producers' price is about 70 eurocent per kilo. If we use as our upper limit that the changes since 2002 are entirely the result of the bird flu outbreak, the total decline in production (270,000 tons in two years' time) can be attributed to bird flu. This represents a value of €189 million for the producer. A proportion of this loss in turnover is already included in the €500 million loss in turnover of eggs and meat. If we assume that 50% of the decline in turnover in 2003 is already included in this figure, the extra damage as a result of the drop in meat and egg production comes to **€200 million**. As our lower limit we will assume that 50% of our drop in turnover is attributable to the bird flu outbreak: **€100 million**.

Table 11.
Estimated costs from the BSE, foot-and-mouth disease, swine fever and bird flu outbreaks in the Netherlands during the past 10 years.

Epizootic	Costs (in millions of euros)	
	Prevention and control	**Losses**
BSE	519-769	35-171
Foot-and-mouth disease:		
Livestock industry (including government & EU funding)	432	233
Tourism, transportation of cargo etc.	-	208
Nature conservation	1	-
Classical swine fever		
Livestock industry (including government & EU funding)	1400	135-462
Bird flu		
Livestock industry (including government & EU funding)	770	100-200
	3122–3372	**711–1274**
Total prevention, control and losses	**3833–4646**	

Summary

Table 11 summarises all estimated costs from the four outbreaks mentioned.

Summarizing, the total costs in the past decade amount to an estimated €3.8 to 4.6 billion. As mentioned, our starting point is that, as a "long-term average", the upper limit is equivalent to the average costs in the period 1995 - 2004, while our lower limit amounts to 25% of that.

We thereby arrive at a figure of **€95–470 million in average annual costs**. The lion's share is for the livestock industry (partially compensated by the national government and the EU): €90-440 million. The remainder is damage to tourism, recreation, retail, construction and transportation: €5-21 million.

Calculating productivity losses from human epidemics

Carin Rougoor PhD
CLM Research and Advice

Wouter van der Weijden MSc
Centre for Agriculture and Environment

Introduction

Diseases generate three types of economic costs:
- Treatment costs
- Prevention and control costs
- Productivity losses

Productivity losses are often less well known than treatment, prevention and control costs. They have to be estimated from the number of sick productive people, the number of working days lost and the value of the lost productivity.

One method of calculating these costs is the 'Human Capital Method'. The number of days not worked due to illness is multiplied by the average salary (income) per day.

Although this method has the charm of simplicity, it has often been criticised for two reasons:
- For short-term diseases like influenza, actual productivity loss is often less than calculated because other employees, or family members take over some of the work of their sick colleagues.
- For long-term (chronic) diseases, the sick employee will often be replaced after a couple of months by taking on a person from the reservoir of unemployed people. Alternatively, someone employed elsewhere is taken on, but then this worker may be replaced by a person from the reservoir of unemployed people. In both cases, overall productivity is restored.

The 'Friction Method' takes these considerations into account by applying a correction factor, e.g. 0.8, to productivity loss. For chronic diseases, it is assumed that the sick person will be replaced after a certain period (e.g. 3 or 6 months) and productivity restored.

However, this method has the disadvantage of being highly dependent on the situation on the labour market. During a period of labour shortage, it will much more difficult to replace a sick skilled worker than under labour surplus conditions. Furthermore, the effects of short-term and long-term diseases may be not as different as assumed. In big companies even short-term diseases affecting 10% of employees will require hiring additional workers. In this report we have combined the two methods, while calculating the productivity losses from flu (short term) and HIV/AIDS (chronic) using a fixed replacement ratio of 60%, corresponding to a correction factor of 0.4.

The influenza virus worldwide

For calculating the global productivity losses from the flu virus, we have made the following assumptions:
- The world population is 6.3 billion.
- On average, 10–15% of the population annually contracts flu.
- Most of those infected are sick for 3 to 7 days. Assuming 5 working days per week, we can roughly translate this into 2 to 5 working days lost. For a multi-year annual global average, we take 3 working days lost as a low estimate and 4 working days lost as a high estimate.
- The global average per capita GDP is $6851 per year.
- The global average number of working days per year is between 250 and 300. Consequently, the average GDP per working day is $27 to $34.
- On average, 60% of lost productive capacity is compensated by colleagues, contracted unemployed workers, or – in the case of self-employed workers such as farmers – by unemployed family members doing the job. The 60% is taken as multi-year global average, ignoring the large differences between regions.

Based on these assumptions, a low estimate is 650 million infected persons x 3 days x $27 x 0.4 = $21 billion per year. A high estimate is 975 million persons x 4 days x $34 x 0.4 = $53 billion per year. **The global productivity losses from the influenza virus are an estimated $21–53 billion per year.**

The influenza virus in the Netherlands

In calculating the productivity losses from the influenza virus in the Netherlands, we made the following assumptions:
- The Dutch population is 16.3 million
- The average number of sick days due to flu per year is 0.25 to 1.0. As a multi-year average we take 0.4–

0.8 days per year.
- Per capita GDP is $38,320 per year. This equals $192 per working day (assuming 200 working days per year).
- On average, 60% of lost capacity is made good by colleagues, hired workers or family members.

Thus a low estimate is 16.3 million persons x 0.4 day x $192 x 0.4 = $501 million per year. A high estimate is 16.3 million persons x 0.8 day x $192 x 0.4 = $1001 million per year. Assuming a conversion ratio of €1 to $1.28, **the productivity losses from the influenza virus in the Netherlands are an estimated €391–782 million per year.**

The HIV virus in the Netherlands

Using the same calculation method once again, we made a rough estimate of the productivity losses from the HIV virus based on the following assumptions:
- The Dutch population is 16.3 million.
- The number of patients is 12,000.
- 50–75% of these have a job (exceeding the average 35% of the whole population, since HIV hits adults rather than children and pensioners).
- On average, HIV-positive employees work 1 to 2 days a week less than before contracting HIV.
- Per capita GDP is $38,320 per year, or $192 per working day.
- On average, 60% of lost capacity is made good by colleagues, hired workers or family members.

Based on these assumptions a low estimate is 12,000 workers x 50%/35% x 1 day/week x 52 weeks x $192 x 0.4 = $68 million per year. Our high estimate is 12,000 x 75%/35% x 2 days/week x 52 weeks x $280 x 0.4 = $205 million per year. Assuming the same conversion ratio of €1 to $1.28, **the productivity losses from the HIV virus are an estimated €53–160 million per year.**

Of course, all these estimates are very rough and some of the assumptions are rather arbitrary. But that may be true of many other estimates as well.

Like every other disease, influenza and AIDS also generate economic winners, such as the healthcare industry and, for example, the "science industry". The same goes for animal diseases (see Box 81) and, in fact, most other problems in society. As in the previous Appendix, we have not included such benefits, since they conceal the problem.

Pathways of species introductions

I.   ***Transportation–Related Pathways*** - *incl. military travel –*

       A.   ***Modes of Transportation*** *(i.e. things doing the transporting)*

            1.   **Air Transportation**

                a)   *Examples - Planes, helicopters, etc. (e.g. stowaways in wheel wells, cargo holds, an anywhere else)*

                    (1)   *Organisms Transported - v (snakes and others), in, inv, ps, pdp - pleas see list at end of this document for abbreviations.*

            2.   **Water/Aquatic Transportation** *- including all methods of moving through the water.*

                a)   *Examples - all types of ships (incl. cruise ships), recreational boats and other craf barges, semi submersible dry docks, oil derricks - can be freshwater or marine c both; can be large or small; includes industrial, tourism, recreational, la enforcement, and military crafts.*

                b)   *Subpathways*

                    (1)   **Ballast Water and Sediments** *and other things that hold water sea chests, engines, etc.*

                        (a)   *Organisms Transported - ai, ap, mbv, di, ph*

                    (2)   **Hull/Surface Fouling**

                        (a)   *Organisms Transported - hfo, other aquatic organism when talking about slow moving platforms*

                    (3)   **Stowaways in holds**, *cabins, etc.*

                        (a)   *Organisms Transported - v, inv, ps, pdp*

                    (4)   **Superstructures/structures above the water line**

                        (a)   *Organisms Transported - inv (gypsy moths), others?*

                    (5)   **Dredge Spoil Material**

                        (a)   *Organisms Transported - ai, av, ap, adp, pdp*

            3.   **Land/Terrestrial Transportation** *- including all methods of moving across the ground.*

                a)   *Subpathways*

                    (1)   *Cars, trucks, buses, ATVs, etc.*

                    (2)   *Construction equipment and firefighting equipment*

                    (3)   *Trains, subways, metros, monorails*

                    (4)   *Hikers, Horses, Pets*

                b)   *Organisms Transported - ps, gm, si, in, v, adp, pdp*

       B.   ***Items Used In Shipping Process***

            Containers *- both exterior and interior*

                a)   *Organisms Transported - ps, gm, si, in, v, dp,*

            Packing Materials

                b)   *Subpathways*

                    (1)   **Wood packing materials** *- wood pallets, wood crates,*

                        (a)   *Organisms Transported - ps, in, pdp, si*

                    (2)   **Seaweed**

                        (a)   *Organisms Transported - ai, av, adp, pdp*

                    (3)   **Other plant materials**

                        (a)   *Organisms Transported - ps, psp, in, si, v, adp*

                    (4)   **Sand/earth** *- sometimes used in archaeological shipments*

                        (a)   *Organisms Transported - in, inv, ps*

       C.   ***Tourism/Travel/Relocation***

            1.   *Examples - travel for recreation, business or for relocation*

            2.   *Subpathways –*

                a)   **Travelers themselves** *(incl. humans as vectors for disease)*

b)   **On baggage and gear** - *"carry on" and checked items*

c)   **Transported Pets/Plants and Animals Transported for Entertainment** – *this includes pets already owned that are transported when one moves or travels, and animals transported for horse shows, sporting events, circuses, rodeos, plant or garden shows, etc.*

d)   **Travel Consumables** *(food on cruise ships, etc.)*

3.   *Organisms Transported - ps, insect, sim inverts, dp*

D.   **Mail/Internet/Overnight Shipping Companies**

1.   *Organisms Transported - ps, pdp, in, si, ai, av*

II.   **Living Industry Pathways** - **[see diagram 3]**

A.   **Food Pathways**

1.   **Live Seafood** *(market ready - imported into and/or throughout the U.S. for immediate consumption)*

a)   *Subpathways*

(1)   *Food organism "in trade" - intentionally released (authorized or unauthorized) or escaped*

(2)   *Hitchhikers*[*]

(a)   *On or in live seafood (incl. parasites and pathogens)*

(b)   *In water, food, packing material, substrate (live rock?)*

b)   *Organisms Transported - ai, ap, av, di, ph, adp, pdp, la*

2.   **Other Live Food Animals** *(imported alive into and/or throughout the U.S.)*

a)   *Examples - Livestock, game birds*

b)   *Subpathways*

(1)   *Food organism "in trade" - intentionally released (authorized or unauthorized) or escaped*

(2)   *Hitchhikers*

(a)   *On or in live animals (incl. parasites and pathogens)*

(b)   *In water, food, growing medium, nesting or bedding*

c)   *Organisms Transported - adp, in, mbv, tv, v*

3.   **Plants and Plant Parts as Food** *(imported into and/or throughout the U.S.)*

a)   *Examples - fruits, vegetables, nuts, roots, seeds, edible flowers, etc.*

b)   *Subpathways*

(1)   *Plant "in trade" - intentionally released (authorized or unauthorized) or escaped*

(2)   *Hitchhikers*

(a)   *On or in food organism (incl. parasites and pathogens)*

(b)   *In water, food, growing medium, nesting or bedding*

c)   *Organisms Transported - ps, pdp, in, inv, v (frogs on plants, etc.)*

B.   **Non-Food Animal Pathways**

1.   **Aquaculture** *(includes the sites where organisms are raised, the raising of the organism, and their movement, unless classified as live seafood; if an organism usually classified as live seafood is being transported for reproduction purposes or other reasons, it falls under aquaculture).*

a)   *Examples - fish, shellfish, shrimp and other invertebrates*

b)   *Subpathways*

(1)   *Aqua cultured organism "in trade" - intentionally released (authorized or unauthorized) or escaped*

(2)   *Hitchhikers*

(a)   *On or in cultured organism (incl. parasites and pathogens)*

(b)   *In water, food, growing medium, nesting or bedding*

c)   *Organisms Transported - when including larval stages of animals, almost any aquatic plant or animal is possible, with the exception of marine mammals*

2. ***Pet/Aquarium Trade*** *- including the organisms and their facilities*

    a)   *Examples - dogs, cats, birds, herpetiles, exotic mammals, fish, other aquarium stock, invertebrates (tarantulas, scorpions, etc.)*

    b)   *Subpathways*

        (1)   *Pet organism "in trade" - intentionally released (authorized or unauthorized) or escaped*

        (2)   *Hitchhikers*

            (a)   *On or with pet organism (incl. parasites and pathogens)*

            (b)   *In water, food, growing medium, nesting or bedding, aquarium substrates,*

    c)   *Organisms Transported - almost anything is possible - see list at end*

3. ***Bait Industry***

    a)   *Examples - anything used as bait for fishing, etc.*

    b)   *Subpathways*

        (1)   *Bait organisms "in trade" - intentionally released (authorized or unauthorized) or escaped*

        (2)   *Hitchhikers*

            (a)   *On or with bait (incl. parasites and pathogens)*

            (b)   *In water, food, growing medium, nesting or bedding*

    c)   *Organisms Transported - ai, ap, av, di, ph, adp, pdp, la*

4. ***Non-Pet Animals***

    a)   *Examples - importation of animals for non-food livestock (hunt clubs, breeding, racing, draft animals), research, harvesting fur/wool/hair, entertainment and their sites of deliberate introduction (zoos, public aquaria, ranches, rodeos, lab facilities, etc.).*

    b)   *Subpathways*

        (1)   *Non-pet organism "in trade" - intentionally released (authorized or unauthorized) or escaped*

        (2)   *Hitchhikers*

            (a)   *On or with non-pet animal (incl. parasites and pathogens)*

            (b)   *In water, food, growing medium, nesting or bedding*

    c)   *Organisms Transported - adp, in, mbv, tv, v*

C.   ***Plant Trade*** *(aquatic and terrestrial)*

    1.   *Examples -importation of plants and sites of deliberate introductions of plants (botanical gardens, nurseries, landscaping facilities, research facilities, public and private plantings, and aquariums/water gardening facilities when talking about aquatics, etc.)*

        a)   ***Whole plants*** *and nurseries/landscaping/garden facilities*

        b)   ***Plant parts***

            (1)   ***Seeds*** *and the seed trade*

            (2)   ***Below Ground Plant Parts***

                (a)   *Bulbs, culms, roots, tubers, etc.*

            (3)   ***Above Ground Plant Parts***

                (a)   *Cuttings, budwood*

            (4)   ***Aquatic Plant Propagules***

    2.   *Subpathways*

        a)   *Plant organisms "in trade" - intentionally released (authorized or unauthorized) or escaped*

        b)   *Hitchhikers*

            (1)   *On or with plant or plant part (incl. parasites and pathogens)*

            (2)   *In water, growing medium, or packing material*

    3   *Organisms Transported - ps, pdp, in, si, v, ai, av, adp*

          Biological globalisation

III.   **Other Miscellaneous Pathways -**
    A.   **Other Aquatic Pathways**
        1.   *Subpathway*
            a)   **Interconnected Waterways**
                (1)   *Examples - Chicago Ship and Sanitary Canal*
            b)   **Inter basin Transfers**
                (1)   *Examples - California Aqueduct, All American Canal*
        2.   *Organisms Transported - ai, av, ap, adp, pdp*
    B.   **Other Animal and Plant Related Pathways**
        1.   **Minimally Processed Animal Products**
            a)   *Examples - hides, trophies, feathers*
            b)   *Organisms Transported - adp, in, inv*
        2.   **Minimally Processed Plant Products**
            c)   *Examples - logs, firewood, chips, mulch, straw, baskets, sod, etc.*
            d)   *Organisms Transported - in, inv, ps, pdp,si, v*
        3.   **Meat Processing Waste**
            a)   *Organisms Transported - adp*
    C.   **Ecosystem Disturbance**
        1)   **Short-term disturbances that facilitate introduction**
            a)   *Examples - habitat creation, restoration, enhancement; forestry*
            b)   *Organisms Transported - ps, pdp, in, inv, v*
        2)   **Long-term disturbances that facilitate introduction**
            a)   *Examples -*
                (1)   *Highway rights-of way, Railroad rights of way, Utility Rights of way*
                (2)   *Land clearing, development, damming, stream channelization, logging*
            b)   *Organisms Transported - ps, pdp, in, inv, v*
    Natural Spread of Established Populations of Invasive Species
        1)   *Examples include natural migration, movement and spread of established populations, ocean currents, wind patterns, unusual weather events, spread by migratory waterfowl, etc.*

Organisms transported - this category includes all established invasive species

Key to Organisms Transported

ai = *aquatic invertebrates (and larval stages)*

adp = *animal disease pathogens and parasites*

ap = *aquatic plants*

av = *aquatic vertebrates (and larval stages)*

di = *dinoflagellates*

dp = *disease pathogens*

gm = *gypsy moth*

hfo = *hull fouling organisms*

in = *insects and similar invertebrates*

inv = *other invertebrates (not insects)*

mbv = *microbes, bacteria, and viruses*

pdp = *plant disease pathogens*

ph = *phytoplankton*

ps = *plants and seeds*

si = *snails and other invertebrates*

tv = *terrestrial invertebrates (insects and other arthropods)*

v = *vertebrates*

`In all places where the term hitchhiker is used, it includes plants, animals, invertebrates, parasites, diseases and pathogens.

Source: Invasive Species Pathways Team (2003)

100 of the world's worst invasive alien species

Disease

avian malaria

banana bunchy top

chestnut blight

crayfish plague

Dutch elm disease

frog chytrid

phytophthora root rot

rinderpest

Aquatic plant

caulerpa seaweed

common cord-grass

wakame seaweed

water hyacinth

Land plant

African tulip tree

black wattle

Brazilian pepper tree

cogon grass

cluster pine

erect pricklypear

fire tree

giant reed

gorse

hiptage

Japanese knotweed

Kahili ginger

Koster's curse

kudzu

lantana

leafy spurge

leucaena

melaleuca

mesquite

miconia

mile-a-minute weed

mimosa

privet

pumpwood

purple loosestrife

quinine tree

shoebutton ardisia

Siam weed

strawberry guava

tamarisk

Micro-organism

Plasmodium relictum

Banana bunchy top virus

Cryphonectria parasitica

Aphanomyces astaci

Ophiostoma ulmi

Batrachochytrium dendrobatidis

Phytophthora cinnamomi

Rinderpest virus

Caulerpa taxifolia

Spartina anglica

Undaria pinnatifida

Eichhornia crassipes

Spathodea campanulata

Acacia mearnsii

Schinus terebinthifolius

Imperata cylindrica

Pinus pinaster

Opuntia stricta

Myrica faya

Arundo donax

Ulex europaeus

Hiptage benghalensis

Polygonum cuspidatum (Fallopia japonica)

Hedychium gardnerianum

Clidemia hirta

Pueraria montana

Lantana camara

Euphorbia esula

Leucaena leucocephala

Melaleuca quinquenervia

Prosopis glandulosa

Miconia calvescens

Mikania micrantha

Mimosa pigra

Ligustrum robustum

Cecropia peltata

Lythrum salicaria

Cinchona pubescens

Ardisia elliptica

Chromolaena odorata

Psidium cattleianum

Tamarix ramosissima

wedelia — *Wedelia trilobata*
yellow Himalayan raspberry — *Rubus ellipticus*

Aquatic invertebrate
Chinese mitten crab — *Eriocheir sinensis*
comb jelly — *Mnemiopsis leidyi*
green crab — *Carcinus maenas*
marine clam — *Potamocorbula amurensis*
Mediterranean mussel — *Mytilus galloprovincialis*
Northern Pacific seastar — *Asterias amurensis*
spiny water flea — *Cercopagis pengoi*
zebra mussel — *Dreissena polymorpha*

Land invertebrate
Argentine ant — *Linepithema humile*
Asian longhorned beetle — *Anoplophora glabripennis*
Asian tiger mosquito — *Aedes albopictus*
big-headed ant — *Pheidole megacephala*
common malaria mosquito — *Anopheles quadrimaculatus*
common wasp — *Vespula vulgaris*
crazy ant — *Anoplolepis gracilipes*
cypress aphid — *Cinara cupressi*
flatworm — *Platydemus manokwari*
Formosan subterranean termite — *Coptotermes formosanus shiraki*
giant African snail — *Achatina fulica*
golden apple snail — *Pomacea canaliculata*
gypsy moth — *Lymantria dispar*
khapra beetle — *Trogoderma granarium*
little fire ant — *Wasmannia auropunctata*
red imported fire ant — *Solenopsis invicta*
rosy wolf snail — *Euglandina rosea*
sweet potato (tobacco) whitefly — *Bemisia tabaci*

Amphibian
bullfrog — *Rana catesbeiana*
cane toad — *Bufo marinus*
Caribbean tree frog — *Eleutherodactylus coqui*

Fish
brown trout — *Salmo trutta*
carp — *Cyprinus carpio*
large-mouth bass — *Micropterus salmoides*
Mozambique tilapia — *Oreochromis mossambicus*
Nile perch — *Lates niloticus*
rainbow trout — *Oncorhynchus mykiss*
walking catfish — *Clarias batrachus*
Western mosquito fish — *Gambusia affinis*

Bird
Indian myna bird — *Acridotheres tristis*
red-vented bulbul — *Pycnonotus cafer*
starling — *Sturnus vulgaris*

Reptile

brown tree snake	*Boiga irregularis*
red-eared slider	*Trachemys scripta*

Mammal

brushtail possum	*Trichosurus vulpecula*
domestic cat	*Felis catus*
goat	*Capra hircus*
grey squirrel	*Sciurus carolinensis*
macaque monkey	*Macaca fascicularis*
mouse	*Mus musculus*
nutria (coypu)	*Myocastor coypus*
pig	*Sus scrofa*
rabbit	*Oryctolagus cuniculus*
red deer	*Cervus elaphus*
red fox	*Vulpes vulpes*
ship rat	*Rattus rattus*
small Indian mongoose	*Herpestes javanicus*
stoat	*Mustela erminea*

Species on the list were selected to illustrate important issues of biological invasion. Absence from the list does not imply that a species poses a lesser threat. The development of the *100 of the World's Worst Invasive Alien Species* list and database has been made possible by the generous contribution of the Fondation d' Entreprise TOTAL. The *Global Invasive Species Database* contains further information on these and other alien invasive species (e.g. ecology, habitat, impacts, pathways of introduction).

Reference: Invasive Species Specialist Group (ISSG) undated: 100 of the World's Worst Invasive Alien Species – A selection from the Gobal Invasive Species Database. www.issg.org/database. Slightly modified.

Glossary of terms

Abiotic: Not living.

Adaptive radiation: A burst of evolution with rapid divergence from a single ancestral species that results from an array of habitats and niches.

Adventitious (adventive) plant: A plant that has been accidentally introduced. Depending on the material with which it was introduced, we can distinguish between grain adventives, wool adventives, stone adventives, etcetera.

Alien species: A species that does not naturally occur in an area where it is found at a particular moment. *Synonyms*: exotic, non-indigenous and non-native species.

Antigenic shift: A genetic shift in a virus, which can make a vaccine or acquired immunity less effective.

Arbovirus: Arthropod Borne Virus, a virus transmitted by arthropods, such as mosquitoes or ticks. Most of these viruses are currently named togaviruses.

Archaeophyte: A plant species that became naturalised in a defined area before a defined year far back in history (often AD 1500 is chosen).

Arthropod borne disease: Disease transmitted by arthropods such as mosquitoes, ticks or fleas.

Biocoenosis: All the organisms living in a particular place at a particular time.

Bio-introduction: Introduction of a biological species by humans, be it intentionally or unintentionally, to an area where it does not yet occur. Not every introduction leads to an invasion nor is every invasion caused by an introduction.

Bio-invasion: The penetration of a species, by whatever cause, into an area new to that species, whereby it maintains itself, reproduces and spreads. Short-lived invasions such as the seasonal invasions of migratory birds and other animals are often not considered bio-invasions proper. Depending on the cause of the invasion, we can distinguish between climate-driven, transport-related, traffic-related, infrastructure-related, habitat change-related invasions, etcetera.

Biological globalisation (biotic globalisation, bio-globalisation): The increasing interchange of plants, animals and micro-organisms between bioregions of the world, leading to biotic homogenisation and numerous new ecological interactions. Insofar as this process is facilitated by man - through transport, traffic, breakdown of distribution barriers and man-made climate change – it can be named **anthropogenic bio-globalisation**, the focus of this book.

Biological proliferation (biotic proliferation, bio-proliferation): The process of biological species expanding their ranges.

Biome: A large ecological community (or complex of communities) that stretches over a large area characterized by a dominant type of vegetation. The boundaries between biomes are partly determined by the climate and may not be sharp. Examples: tundra, taiga, tropical rainforest and desert.

Bioregion (biogeographical region): A region with a relatively homogeneous flora and fauna, more or less separated from other such regions by geographical barriers.

Biota: The total of species in a particular region at a particular time. Simplified: the flora and fauna together. See also: community.

Bio-terrorism: Terrorism applying biological weapons against people, livestock or crops.

Biotic: Living.

Biotic homogenisation: The mixing of biotas (in this book: on a global scale) leading to increasing similarity and decreasing variation between them. Biodiversity increases locally, at least in early stages, but often declines globally, with some successful species expanding their ranges and other, less competitive species declining or going extinct.

Biotic resistance: Regulation of exotic species by native natural enemies.

Case-fatality rate: The deceased fraction of a given number of patients with the same disease, varying from 0% to 100%.

Colonization: The expansion of a species into an area new for that species. Successful colonization leads to settlement (establishment).

Commensalism: Mutualism between two species where one species benefits and the other is not harmed.

Community: An assemblage of plant and animal species living in a particular area or habitat at a particular time. Communities are often named after one of their dominant species (e.g. a pine community) or the major physical characteristics of the area (e.g. a freshwater pond community). Members interact in various ways, including food chains and competition.

Counter-invasion *(in this book)*: Invasion of a natural enemy of an invasive pest species. Bright (1998): the attack on an exotic species by an indigenous species. However, this is no proper use of the word "invasion".

Counter-pest (Elton 1958): A natural enemy of a pest species introduced to control that species.

Endemic disease *(in epidemiology)*: A disease that occurs continuously in a certain area, at a more or less constant level.

Endemic species *(in biogeography)*: A species occurring exclusively in a certain area.

Epidemic: A significant increase, over a short period of time, of the incidence of a certain disease in humans. Other than is often assumed, this applies *not* only to infectious diseases but also to other diseases such as cardiovascular diseases.

Epizootic: An epidemic in an animal population.

Exotic species: A species that does not naturally occur in the area where it is found at a particular moment. *Synonyms*: alien, non-indigenous and non-native species.

Fomite: Dead object that serves as a vehicle for pathogens. Examples: razor (hepatitis B), drinking glass (glandular fever) and faucet (influenza).

Globalisation: The increasing interchange of goods, services, information, organisms, people and money between regions of the world. The main types of globalisation are **economic, socio-cultural** and **biological** globalisation.

Import: The import (by humans) of a species into an area. This can either be a species that is new to that area, or a species that is native to it. Examples: salmon and oysters. Imported organisms can differ genetically from the population already present in the area.

Import-linked cases *(in epidemiology)*: Cases of infectious diseases that have been brought into a country by patients. Examples: malaria, typhoid fever, hepatitis A and infections by the MRSA bacterium.

Incidence: The number of *new* cases of a certain disease in a well-defined period of time (often one year) in a defined population.

Indigenous species: A species that occurs in a defined area at least since a defined year in the past. *Synonym*: native species.

Introduction: The transport of a species into an area where it does not yet occur. This can occur intentionally, unintentionally, or intentionally but consciously.

Invasion: The settlement and subsequent spreading of a species in a new area.

Invasive species: At least five different definitions of this term are given in the literature (Colautti & MacIsaac 2004). *In this book:* A species that colonizes a new area, maintains itself, reproduces and spreads spontaneously. According to this definition, an invasive species is *not* necessarily *wide*spread or harmful in its new area. For comparison: the definition of invasive species used by the US federal government is limited to species that are harmful or likely to cause harm.

Isopoda: order of Crustacea, including wood-lice, water-slaters, pillbugs etc.

Lag phase: The phase in which a species has settled in an area but has not yet significantly expanded there. This phase can last more than 100 years.

Monophagous: Feeding on only one type of plant or prey or on species that are closely related.

Native species: A species that for at least millennia occurs in a particular area. *Synonym*: indigenous species.

Naturalisation: Settling by a species in an area, reproducing and maintaining or increasing its numbers at more than one location during at least a specified unit of time or number of generations.

Neophyte: A plant species that became naturalised in a defined area after a defined year not too far back in history, e.g. AD 1500.

Niche: The status or role or a species in a habitat, defined in terms of food, predators, temperature tolerance etc. Gause's principle states that two species cannot occupy the same niche for an extended period of time.

Oligophagous: Feeding on few types of plant or prey.

Oomycota: a phylum formely classified as fungi (Oomycetes). Their cell wall is made of cellulose and they can reproduce both sexually and asexually. They include potato blight *(Phytophthora)*.

Pandemic: A globally or almost globally occurring epidemic; can stretch over a period of many years.

Panzootic: A globally or almost globally occurring epizootic; can stretch over a period of many years.

Polychaeta: a class of aquatic annelid worms in which each body segment has a pair of flattened fleshy lobes bearing numerous bristles *(chaetae)*.

Polyphagous: Feeding on many types of plants or prey.

Ponto-Caspian species: Species originating in the Black Sea or the Caspian Sea.

Presence: The presence of a species, regardless of the number of individuals found, measured by the number of locations at which it is found and/or the number of measuring points over time where it is found.

Prevalence: The *total* number of cases of a particular disease during a well-defined time period (mostly one year; **annual prevalence**) or at a certain point in time (**point prevalence**) in a defined population (often expressed per 100,000). So cases that had developed *before* the measured period or point in time are also included. In the case of chronic diseases the (point) prevalence can be many times the annual incidence.

Prion: Short for **pro**teinaceous **in**fectious particle. Abnormally structured form of a host protein, able to convert normal molecules of the protein into the abnormal structure. A prion is a unique type of infectious agent, being made only of protein without DNA.

Propagule (invasion) pressure: The number of individuals (or parts, seeds, spores, eggs) of a species that are involved in an arrival in a new area.

Reintroduction: The introduction of a species in an area where it had previously become extinct, with the intention of enabling that species to settle and spread in that area.

Rickettsia: A very small bacterium belonging to the phylum of Proteobacteria. Rickettsias can infect ticks, fleas, lice and mites, through which they can be transmitted to vertebrates, including humans. The group includes the causal agents of Q fever, Rocky Mountain spotted fever and forms of typhus.

Secondary bio-invasion: Invasion of a species not directly from the species' native area of distribution, but from an intermediate area colonized earlier.

Settlement: Establishment of a (non-indigenous) species in a particular area following colonization.

Sylvatic *(in epidemiology)*: Term referring to wild, or forest-dwelling animals. E.g. foxes may be the 'sylvatic reservoir' for the rabies virus, burrowing rodents for the plague bacterium, and primates for the yellow fever virus.

Swamping: The overwhelming of an area, including its inhabitants, with large numbers of an invasive species.

Tariff escalation: A stepwise increase in import duties whereby the tariff rises as the product has been processed further.

Taxon: Any named group of any rank in the hierarchical classification of organisms. E.g. *Sus* or pig (genus), Papilionidae or butterflies (family), Aves or birds (order), and Chordata (a phylum that includes vertebrates). Subspecific taxa include subspecies, varieties, cultivars, strains, lineages and sublineages.

Transforming species: A species that radically changes an ecosystem, either directly or indirectly. For example: rabbits inhibiting the regeneration of forest; or even: a virus that eradicates rabbits, allowing the forest to regenerate.

Tunicata: A subphylum of marine nonvertebrate chordates in which the body is enclosed in a protective *tunic* of a cellulose-like material.

Vector *(in ecology)*: A species' means of transport or carrier to the destination area.

Vector *(in epidemiology)*: An organism transmitting a pathogen from one host to another (animal-to-animal, animal-to-human, human-to-human, human-to-animal). Often this involves arthropod-borne diseases. See also: **fomite**.

Vector-borne disease: Disease transmitted by a vector (mostly insects or ticks) (as opposed to directly transmitted diseases and to diseases transmitted by a fomite).

Zoonosis: A disease transmitted from animals to humans.

Index*

     Biological globalisation

Acknowledgments

During the process of writing this book, we received valuable support from many people. Special thanks are due to:

- Deniz Haydar MSc, Centre for Ecological and Evolutionary Studies, University of Groningen, for valuable information and advice.
- Carin Rougoor PhD, CLM Research and Advice, for calculations of the economic costs from animal and human diseases.
- Peter Boomgaard PhD, professor of Economic and Environmental History of Southeast Asia, University of Amsterdam, for valuable information and comments on some historical issues.
- Hans Schinkel, biologist, Assen, for comments on the complete draft text, for taking the initiative to rewrite and publish our report in book form and taking some photos.
- Paul Kemmeren and Wilma Seijbel of KNNV publishing, for their support for the book edition and for valuable suggestions.

We have also received valuable information, comments and/or suggestions from:

- Jaap Bakker, professor of Nematology, Wageningen University.
- Auke Bijlsma MSc, biologist, Netherlands Organisation for Scientific Research (NWO).
- Floris Brekelmans MSc, *Bureau Stadsnatuur*, City of Rotterdam.
- Rein Cremer MSc, biologist.
- Kees-Jaap Hin, MSc, CLM Research and Advice, Culemborg (now at Schuttelaar and Partners, The Hague).
- Guus de Hollander PhD, National Institute of Public Health and Environmental Protection (RIVM), Bilthoven.
- Bart G.J. Knols PhD, medical entomologist, Laboratory of Entomology, Wageningen University.
- Jan Lever PhD, professor emeritus of Zoology, Free University, Amsterdam.
- Ruud van der Meijden PhD, *Nationaal Herbarium Nederland* (National Herbarium of the Netherlands) of Leiden University.
- Johan Polder PhD, National Institute of Public Health and Environmental Protection (RIVM), Bilthoven.
- Herbert Prins PhD, professor of Nature Conservancy in the Tropics and the Ecology of Vertebrates, Wageningen University.
- Aad Smaal PhD, Centre for Shellfish Research, Yerseke.
- Laurens C. Smits MSc, Plant Protection Service, Wageningen.
- Willem Takken PhD, associate professor of Entomology, Wageningen University.
- Carla Teune, curator emeritus of the Botanical Garden of Leiden University.
- Henk van der Weijden BSc, State Forestry Service (retired).
- Frank de Wolf MD PhD, HIV Monitoring Foundation, Amsterdam.

In addition, Folchert R. van Dijken PhD and Jacob-Jan Bakker MSc of the Directorate for Nature at the Netherlands Ministry of Agriculture, Nature and Food Quality also made valuable suggestions.

Jorrit van Hertum of ETI Bioinformatics translated most of our draft in English. Derek Middleton improved our English in some other paragraphs and boxes.

Bea de Groot carefully checked the list of references, Erik van Well and Francien de Groot made some of the maps and Lies Rookhuizen took some pictures.

Joe LaForest of The Bugwood Network, University of Georgia, assisted in obtaining copyrights of many photos

Last but not least, we wish to thank – some posthumously – a few of our respective mentors, who inspired our studies of biological globalisation:

- Kees den Hartog PhD, professor emeritus of Aquatic Ecology, Catholic University of Nijmegen (now the Radboud University).
- Joop Huisman MD PhD, professor emeritus of Epidemiology, Erasmus University (Rotterdam) and Delft University of Technology.
- Evert Timmer MD, medical epidemiologist, University of Amsterdam (deceased 2003).
- Karel H. Voous PhD, professor emeritus of Systematics and Geography of Animals, Free University, Amsterdam (deceased 2002).

The authors

About the authors

Wouter van der Weijden (1946) studied biology at the Vrije Universiteit Amsterdam and Leiden University, specialising in the taxonomy and geography of animals and in terrestrial ecology. He has published several articles in ornithological journals on the taxonomy of owls and on the birds of Amsterdam. At Leiden University he worked on agriculture and nature issues, later broadening his interest to include sustainable agriculture, on which he co-authored a report for the Scientific Council for Government Policy (1984). In 1980 he co-founded the Centre for Agriculture and Environment and is the Centre's co-director. He conducted research and provides advice on agri-environmental issues, such as farmland biodiversity, on-farm nutrient management, sustainable agricultural and trade policy and sustainable food chains. He has published hundreds of articles and columns, and recently co-authored the book *Agrarian Landscapes across Europe and North America*. In 2002, Wouter rekindled his original passion for biogeography and started work on bioinvasions. His other passions include bird watching.
E-mail: wvanderweijden@clm.nl

Rob Leewis (1947) studied biology, later specialising in marine biology and ecology. He wrote a PhD thesis on the phytoplankton off the Dutch coast (1985). He has worked as a project manager with *Rijkswaterstaat*, the National Water Authority, on the environmental impacts of the Delta Works in the southwest of the Netherlands. He studied underwater life on dikes and shipwrecks, and co-founded the European Artificial Reef Research Network. He also participated in several underwater archaeological projects. Later he worked at the National Institute of Public Health and Environmental Protection (RIVM), coordinating marine research to inform and advice the government on various environmental issues. Currently Rob is chairman of the Dutch Working Group on Exotic Species and a guest researcher for exotic species at the National Museum of Natural History (Naturalis). He is a professional diver and underwater photographer, has produced several books and CD-ROMs on underwater life, and writes for a sport diver magazine.
E-mail: r.leewis@casema.nl

Pieter Bol (1948) began studying archaeology and prehistory, but switched to medicine. After obtaining his MD he conducted research into the bacteriology and epidemiology of meningitis (PhD thesis 1987). He was then secretary to the Dutch Health Council for seven years. Since 1993 he has been associate professor of Public Hygiene and Epidemiology at Delft University of Technology. Pieter is a freelance writer and is the author of hundreds of articles on medicine, the environment and history.
E-mail: pbol@xs4all.nl

Colophon

Authors: Wouter van der Weijden, Rob Leewis, Pieter Bol

Translation: Jorrit van Hertum, ETI, University of Amsterdam, the Netherlands, and Derek Middleton, the Netherlands

Graphic design and production:
Erik de Bruin, Varwig Design, the Netherlands

Picture editing: Weia Reinboud

Printer: DZS, Ljubljana, Slovenia

This publication was supported financially by:
Foundation Funds KNNV

The Dutch Ministry of Agriculture, Nature and Food Quality

Innovation Fund of the Centre for Agriculture and
Environment Foundation, with general donations from Campina,
Cosun, LLTB, LTO-Noord, Rabobank, Product Board for Horticulture,
Triodos Bank and ZLTO

© KNNV Publishing, Utrecht, the Netherlands, and
W.J. van der Weijden, R. Leewis and P. Bol, 2007

The moral rights of the authors have been asserted.

First published February 2007

KNNV Publishing is a foundation of the Royal Dutch Society for Natural History
ISBN 978-90-5011-243-7
www.knnvpublishing.nl